The Wincanton district — a concise account of t

The Wincanton area lies at the north-western end of the Wessex Basin and south-east of the Mendips. The district is mostly rural, with the small towns of Wincanton, Gillingham, Mere and Bruton the principal urban areas. The contrasting scenery reflects the underlying geology. Successive west-facing escarpments are formed by the Inferior Oolite, Fuller's Earth Rock, Forest Marble, Upper Greensand and Chalk; the prominent escarpments of the last two dominate the skyline.

The geological sequence ranges from low in the Lower Jurassic to the Upper Cretaceous Chalk. Details of the outcrops of these strata are presented; they incorporate much new stratigraphical and palaeontological data. Deep boreholes and geophysical data have been interpreted to provide descriptions of the concealed formations and also to elucidate the structure. Extensive areas of landslip have been mapped below the escarpments of the Corallian Group and the Upper Greensand.

A summary of all the groups, formations and members that appear on the geological map, together with a lithological summary and thickness, is presented in Table 1. Formations below the Blue Lias (Lias Group) are not exposed in the district, and are encountered only in boreholes. Thickness of the formations below the Blue Lias are those proved in the Norton Ferris and Bruton boreholes.

Chapter One gives a résumé of the geological history of the district. Chapter Two, Applied Geology, deals with issues such as the mineral and water resources, landfill and waste disposal, foundation conditions, landslip and ground stability, undermining and solution subsidence, made and worked ground, radon and conservation.

Chapter Three provides an account of the concealed formations beneath the district. These range from the Silurian to the Lower Jurassic Blue Lias. A new stratigraphy for the Lias Group of this part of Somerset is detailed in Chapter Four. Chapters Five and Six cover the rest of the Jurassic System (Inferior Oolite to Purbeck Formation). Previously unrecognised in the district, the outcrop of the Corallian Group has been mapped north of the Mere Fault where it is overlain unconformably by Cretaceous strata. For the first time, important detail from several localities in the poorly exposed mudstone formations of the Oxford Clay and Kimmeridge Clay has been gathered. Details of the Portland Group include photomicrographs of representative lithologies.

Cretaceous strata, ranging in age for the Aptian (Lower Greensand) to the Coniacian (Seaford Chalk), are described in Chapter Seven. The richly fossiliferous Cenomanian part of the Upper Greensand, the Melbury Sandstone, is discussed in detail.

The Mere Fault and associated Wardour Monocline are the major structures of the district. These are documented in Chapter Eight, together with other important faults.

Quaternary deposits include residual Clay-with-flints developed on the Chalk, and the floodplain alluvium; these are described briefly in Chapter Nine.

The section on *Information Sources* details the wealth of data held by BGS, both published and unpublished. The memoir concludes with a comprehensive reference list. Throughout, there is a lavish use of figures, including isopachyte maps and depth-converted maps derived from seismic data, plates and tables.

Cover photograph

Old Wardour Castle. The 15th century castle, the former seat of the earls of Arundel, is largely built out of the local Tisbury Stone of Portlandian age; blocks of the Upper Greensand Shaftesbury Sandstone are also built into the lower part. (Photo: English Heritage K930287)

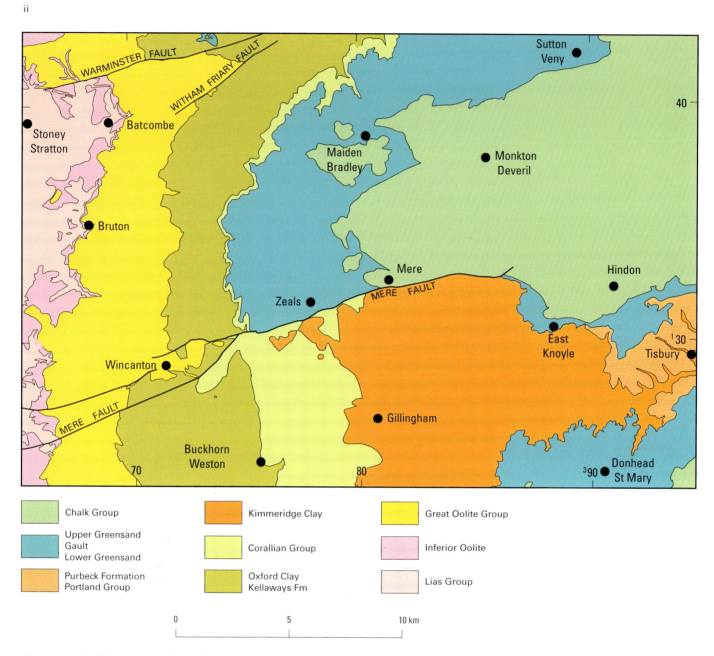

Figure 1 Solid geology of the district.

C R BRISTOW
C M BARTON
R K WESTHEAD
E C FRESHNEY
B M COX and
M A WOODS

The Wincanton district — a concise account of the geology

Memoir for 1:50 000 Geological Sheet 297
(England and Wales)

CONTRIBUTORS

Biostratigraphy and sedimentology
D T Donovan
H C Ivimey-Cook
G K Lott
J B Riding
I P Wilkinson

Deep geology and basin analysis
D J Evans
G A Kirby

Geochemistry
T K Ball

Geophysics
J D Cornwell
B C Chacksfield

Hydrogeology
C S Cheney

London: The Stationery Office   1999

ISBN 0 11 884551 9

Bibliographical reference

BRISTOW, C R, BARTON, C M, WESTHEAD, R K, FRESHNEY, E C, COX, B M, and WOODS, M A. 1999.   The Wincanton district — a concise account of the geology.   *Memoir of the British Geological Survey*, Sheet 297 (England and Wales)

Authors

C R Bristow, BSc, PhD
C M Barton, BSc, PhD
British Geological Survey, Exeter

E C Freshney, BSc, PhD
formerly *British Geological Survey*

B M Cox, BSc, PhD
R K Westhead, BSc, PhD
M A Woods, Bsc
British Geological Survey, Keyworth

Contributors

T K Ball, BSc, PhD
B C Chacksfield, BSc
J D Cornwell, BSc, PhD
D J Evans, BSc, PhD
G A Kirby, BSc, PhD
G K Lott, BSc, PhD
J B Riding, BSc, PhD
I P Wilkinson, BSc, PhD
British Geological Survey, Keyworth

H C Ivimey-Cook, BSc, PhD
formerly *British Geological Survey*

C S Cheney, BSc
British Geological Survey, Wallingford

D T Donovan, BSc, PhD
PHG Consulting, London

Printed in the UK for The Stationery Office

J98391 C6   11/99

CONTENTS

FIGURES

PLATES

TABLES

PREFACE

The area around Wincanton has long been a source of inspiration to writers; for example, it provided the setting for much of Thomas Hardy's *Jude the Obscure*. It has also been an area of great interest to geologists since the Georgian days of Miss Ethelred Benett (1746–1845); her book, '*A Catalogue of the Organic Remains of the County of Wilts*' was published in 1831.

This district, together with the adjacent Shaftesbury district to the south, forms part of the Wessex Basin Project. The new survey of the area has involved a comprehensive stratigraphical synthesis of the Jurassic, Cretaceous and Quaternary strata. Major revisions of the Lias Group, Oxford Clay, Corallian Group, Portland Group, Upper Greensand and Chalk are presented.

The detailed maps will be particularly relevant to the needs of the extractive industry, the water supply authorities, planners and civil engineers. For example, extensive landslips have been recognised and delineated for the first time. These are developed particularly around the northern outskirts of the expanding town of Shaftesbury, and form a significant constraint to development in certain areas.

Boreholes drilled for BGS as part of the study provide a wealth of data in areas of poorly exposed strata, for which there was little previous information. Geophysical logs from these and other shallow boreholes provide a good correlation with commercial boreholes, drilled for water and hydrocarbon exploration.

A structural synthesis combines evidence from field mapping with seismic data generated by various oil companies. It provides evidence for both the small- and large-scale structures that have controlled the evolution of the Wessex Basin, and are relevant to the study of hydrocarbon generation and entrapment.

I hope that this new and fascinating account of the geology of the Wincanton area is read not only by geologists, but also by the many people who are interested in better understanding the geology, the landscape and the evolution of the Wessex Basin.

David A Falvey, PhD
Director

British Geological Survey
Kingsley Dunham Centre
Keyworth Nottingham
NG12 5GG

Acknowledgements

The authors thank the following for their assistance: BHP Petroleum Ltd and Kelt UK Ltd for permission to use unpublished oil-company data: Prof. J H Callomon for details of temporary sections: Mr T Cosgrove, formerly of Wessex Water, for copies of geophysical logs in the Font-hill Bishop area: the Dewey Museum, Warminster, for the loan of specimens: Dorset and Wiltshire County Councils for borehole logs and trial pit records for the Gillingham Inner Relief Road and the East Knoyle Bypass, respectively: Mr M Edmunds for details of temporary sections: English Nature (both the Wiltshire, and Somerset & Avon teams) for details of the SSSIs in the district: Mr R Green of Rust Consulting Ltd for copies of the Wincanton Bypass site-investigation reports: Miss C M Hebditch of the Dorset County Museum for help with early references: Mr S Hookins for the donation of Kimmeridge Clay fossils: Mrs L Light of Gillingham Museum for the loan of specimens and details of wells in the Gillingham area: Mr H C Prudden for details of temporary sections and donation of many specimens: Mrs M S Ross for information of some of the brick pits in the district: Saxton Deep Drillers for permission to geophysically log the Bayford Borehole: Mr D Sole for the donation of Oxfordian and Kimmeridgian fossils: Dr V Stevens for help in the field: Mr A Walker for the loan of an ammonite from the Chalk: Mr C J Wood for helpful discussions on Chalk biostratigraphy.

Mrs M Mann of Newpark Gate, Bratton Seymour, and Mr D L James of the Highlands, Batcombe, allowed BGS to drill stratigraphical boreholes, respectively Bratton Seymour and Lodge Farm, on their ground.

The memoir has been compiled by C R Bristow from contributions by the following authors, each of whom is primarily responsible for the sections included in parentheses, but who have also contributed to other sections of the memoir.

One   Introduction
C R Bristow

Two   Applied geology
C R Bristow, C S Cheney (water resources), T K Ball (radon)

Three   Concealed geology
C M Barton, D J Evans and G A Kirby, with J D Cornwell and B C Chacksfield (geophysical interpretations)

Four   Lower Jurassic
R K Westhead with D T Donovan

Five   Middle Jurassic
R K Westhead (Inferior Oolite), E C Freshney (Fuller's Earth, Frome Clay), C M Barton (Forest Marble, Cornbrash)

Six   Middle to Upper Jurassic
B M Cox (Oxford Clay, with R K Westhead; Kimmeridge Clay), C R Bristow (Corallian Group, Portland Group)

Seven   Cretaceous
C R Bristow (Lower Greensand, Gault; Upper Greensand and Chalk with M A Woods)

Eight   Structure
C M Barton, D J Evans and G A Kirby

Nine   Quaternary
C R Bristow

Information sources   C R Bristow

Other contributors include H C Ivimey-Cook (Lower and Middle Jurassic biostratigraphy), G K Lott (thin-section petrography), J B Riding (Jurassic and Cretaceous palynology), I P Wilkinson (Jurassic and Cretaceous calcareous microfauna).

The memoir was edited by M G Sumbler and A A Jackson.

Notes

The word 'district' used in this memoir means the area included in 1:50 000 Series Sheet 297 Wincanton.

Much of the information in this memoir is taken from Open-file Reports listed in Information Sources (p.92).

Boreholes cited in the text are also listed in Information Sources (p.93).

Numbers prefixed by E refer to specimens in the English Sliced Rock Collection of the British Geological Survey.

Numbers preceded by the letter A refer to photographs in the Geological Survey collections.

The authorship of fossil species is given in the inventory of fossils.

Figures in square brackets are National Grid references; places within the Wincanton district lie within the 100 km square ST.

x

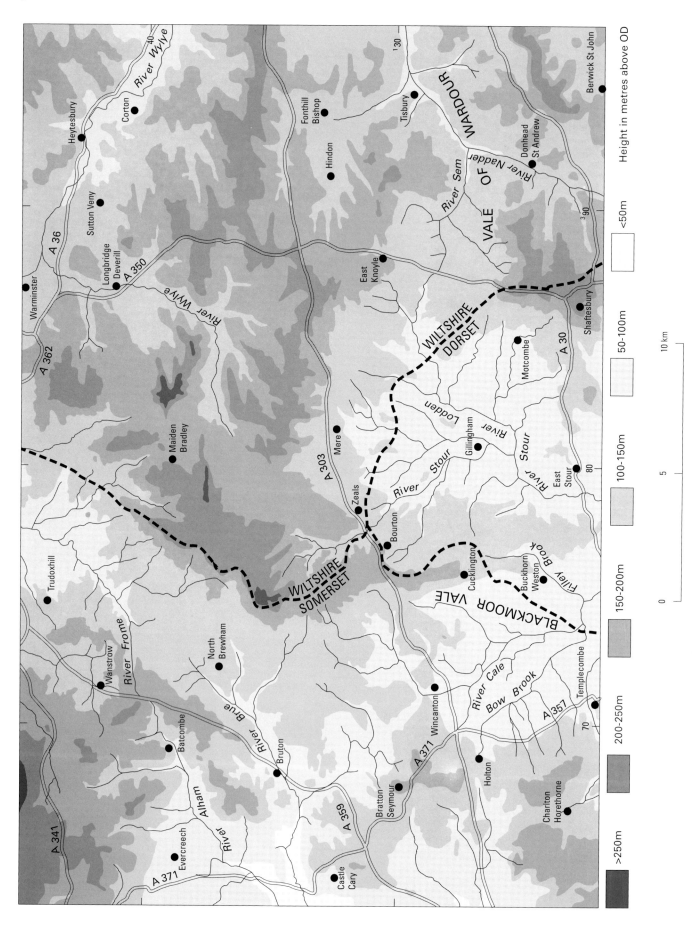

Figure 2 Physical geography of the district and the adjoining areas.

ONE

Introduction

The Wincanton district is mainly rural and falls in the counties of Wiltshire, Somerset and Dorset. The district takes its name from Wincanton (pop. 4700) in Somerset, formerly a woollen trading town and important staging post on the road to London.

The geological deposits of the district are listed in Table 1 (inside the front cover of this memoir). The oldest rocks to crop out, the Jurassic Blue Lias, occur over a small area near Stoney Stratton in the north-west; younger Jurassic strata crop out successively from west to east, with the youngest, the Purbeck Formation, in the extreme east. Cretaceous deposits comprise the Lower Greensand, only locally developed in the east, Gault, with its narrow, heavily landslipped crop, and Upper Greensand and Chalk with their wide outcrops principally in the north-east (Figure 1). At depth, Triassic, Permian and Carboniferous formations have been proved in boreholes drilled for hydrocarbon exploration. For the most part, drift deposits are not widespread. Landslips, however, are extensive in the centre and south-east of the district, where they are associated with the Gault, Oxford Clay and Hazelbury Bryan Formation.

The topography is varied and related directly to the geology (Figure 2). In the west, there is an eastwards succession of prominent, westward-facing escarpments, commonly with long east-facing dip slopes formed by the Inferior Oolite, Fuller's Earth Rock, Forest Marble, Cucklington Oolite (Corallian Group), and Portland Group. These rise above the clay vales; in the south, one of the broadest, the Blackmoor (or Blackmore) Vale, is developed principally on Oxford Clay, although the name is also loosely applied to the broad vale of Kimmeridge Clay. Scarps formed by the Upper Greensand and Upper Chalk have a more varied aspect and curve around a broad syncline in the north-east of the district. These escarpments rise to maximum heights, respectively, of 260 m and 288 m above Ordnance Datum (OD), and dominate the scenery of the centre of the district, so that drainage radiates from this central high ground. The low-lying areas, drained by numerous streams and rivers, were formerly swampy and devoid of settlement. Drainage works have improved the boggy conditions, but the area is still only sparsely populated. Most of the principal settlements are sited on the well-drained limestone, chalk and sandstone units. Gillingham, on Kimmeridge Clay, is an exception to this settlement pattern.

Palaeozoic rocks, which form the sub-Mesozoic basement, include Old Red Sandstone and Carboniferous Limestone (Table 1). The latter locally formed 'highs' during the Mesozoic Era, over which the sediments thin. A period of crustal subsidence that began in the Early Permian initiated the formation of the Wessex Basin. The district lies in the north-west of this major basin. It is crossed by east–west-trending fault zones which subdivide it into smaller basins and highs, of which the Mere Basin is the most important in the district.

The Mere Basin is bounded to the north by the Mendip Hills and to the south by the Cranborne–Fordingbridge High. It controlled deposition during Triassic time, but was less important during the Jurassic, and seems to have no effect on deposition patterns during the Cretaceous Period. The basal Permo-Triassic deposits are about 350 m thick in the centre of the basin. However, they are absent at crop to the north (in the Frome district), in the Bruton and Norton Ferris boreholes in this district and in the Fifehead Magdalen borehole to the south. The overlying Mercia Mudstone Group is up to 400 m thick in the basin centre, but thins onto the Bruton, Fifehead and Mendip highs. The group consists dominantly of mudstones laid down under continental red-bed fluvial conditions. At the close of the Triassic Period, marine conditions returned with the deposition of the Westbury Formation shales and continued throughout much of the Mesozoic Era.

With the exception of the deposition of the Purbeck Formation in the Late Jurassic, fully marine conditions existed across the district throughout Jurassic times. In the Early Jurassic, deposition was more-or-less continuous, but there is evidence of minor erosional or non-depositional breaks in the Lias Group in the north-west of the district, close to a shoreline along the Mendip High. There was a break in deposition in Middle Jurassic times. Across much of the district, Aalenian Upper Inferior Oolite rests unconformably on Toarcian Bridport Sands. Just north of the district, the Inferior Oolite oversteps all Jurassic formations to rest on Carboniferous Limestone. This was probably the first of the Mesozoic formations to extend across the Mendip High. The succeeding Great Oolite Group is represented mostly by mudstones laid down in a quiet, low-energy environment. The Cornbrash is a widespread marine limestone divisible into lower (Bathonian) and upper (Callovian) units, generally separated by a non-sequence. The succeeding shallow marine, sandy mudstone of the Kellaways Formation pass up into the Oxford Clay Formation, a sequence of shallow-water, marine-shelf mudstones overlain in turn, by mudstones, sandstones and limestones of the Corallian Group that form a series of clastic/carbonate shallowing-upward cycles. The Kimmeridge Clay was deposited during a period of high global sea level in a low-energy environment that varied from aerobic to anaerobic.

Towards the end of Kimmeridgian times, the sea became shallower, and the Portlandian deposits are interpreted as shallow-water marine deposits. Further shallowing, accompanied by a probable north-eastwards retreat of the sea during the Late Portlandian, established a lagoonal environment in the east of the district, in which the Purbeck Formation was deposited.

After Ryazanian times, there was uplift and erosion of the western part of the Wessex Basin. Erosion was not uniform across the district, but was probably most severe on the northern flank of the Mere Basin. In the north and west, any Berriasian, Portlandian and Kimmeridgian deposits were removed during a period of erosion associated with the 'Late-Cimmerian Orogeny'. This is represented by a series of breaks in Late Jurassic and Early Cretaceous sequences across north-west Europe.

The Lower Greensand (Aptian to possibly Early Albian) was deposited during the first pulse of a marine transgression. Succeeding beds overstep westwards across an eroded Jurassic surface. The Lower Greensand is overlapped by the Gault (Middle Albian), the first Cretaceous formation to extend across the district. Deposition during the rest of the Cretaceous was fairly uniform, but hardgrounds and glauconitised nodular beds in the chalk indicate local shallowing and erosion.

At the close of the Cretaceous Period, uplift and minor folding were followed by erosion. No Palaeogene deposit is preserved in the district, but at the end of this period, a widespread phase of folding and faulting, the Alpine Orogeny, culminated in the contraction of the major Mesozoic basins and highs. The last movements along the major faults in the district probably took place during the Miocene. No Neogene deposit is known.

Clay-with-flints probably formed during the Pleistocene Period. The formation of head is thought to have commenced under freeze/thaw conditions; some accretion continues by colluvial process at the present day. In the late Pleistocene, probably during the Devensian cold stage, meltwater from semipermanent snow caps on the higher ground transported large quantities of sand and gravel and deposited them as river terrace deposits. Landslips were probably initiated at that time; some movement still occurs. Deposition of the alluvium, initiated during the Holocene, continues to the present day.

TWO

Applied geology

In this chapter, the geological factors relevant to land-use planning and development in the district are reviewed. The key issues are identified; some are considered in detail, but in all cases, the readers should carry out their own investigations before undertaking any development.

KEY ISSUES

Despite the district's rural nature, there has been a long history of small- to medium-scale quarrying and pitting from most formations. The only major developments were a brick pit at Gillingham, now closed, and Charnage Down limeworks. Because of the relative small scale of these operations, there is no legacy of extensive contaminated and undermined ground, but there still remain many geologically related issues which need to be taken into account in land-use planning and development. These are listed below.

- *Mineral resources*: past exploitation, remaining resources, current activity, sterilisation under new development.
- *Water resources*: surface water, groundwater, aquifer vulnerability and pollution.
- *Landfill and waste disposal.*
- *Foundation conditions*: geotechnical properties of rocks and soils, weathering of mudstones, differential settlement, shrink-swell clays, compressible organic deposits.
- *Landslips and ground stability*: past and potential mass movements.
- *Undermining, solution-subsidence and collapse structures.*
- *Made and worked ground*: extent, compositional variation, industrial contamination, toxic residues, leachate mobility.
- *Radon*: potentially hazardous high levels.
- *Conservation*: protection of sites of scientific interest.

MINERAL RESOURCES

The following minerals have been worked in the district: clay for bricks, tiles and pipes: limestone and sandstone for building, wall stones and tiles: chalk and limestone for agricultural lime: fine-grained, siliceous sandstone for millstones and scythestones: sandstone, limestone, chalkstone and, locally, gravel for road aggregate, and sand for building sand.

The outcrops of most potentially workable deposits are widespread in the district, and sterilisation by new developments will only have a local effect. Details of the many old pits and quarries can be found in the Technical Reports listed in Information sources (p.92).

Brick clay

The following formations and members have been dug for brick clay: **Frome Clay** north-north-east of Maperton: **Kellaways Formation** north of Wanstrow: **Mohuns Park Member** at Bathpool Lane, Horsington Marsh: the bituminous **Peterborough Member** and the lower part of the **Stewartby and Weymouth members (undivided)** at Brick Yard Belt north-east of Witham Hall Farm and at Kington Magna (up to the 1920s) where bricks, tiles and flower pots were made (Ross, 1985; Young, 1972): **Hazelbury Bryan Formation** for bricks, just west of Witham Park: **Kimmeridge Clay** at Bourton, Sedgehill, East Knoyle, Hawker's Hill, Motcombe (in use to 1939) (Young, 1972), King's Court, Gillingham (Ross, 1992), and at Gillingham where two areas were worked by the Gillingham Pottery, Brick and Tile Company, the largest and last company to operate (1866–1969). The first, west of the B3092, worked Lower Kimmeridge Clay and **'Ampthill Clay'**. This pit has now been partially backfilled, levelled and turned into an industrial estate. A later pit, east of the road, is flooded. The company made extruded wire-cut bricks, tiles, drain pipes and terracotta goods (Young, 1972). A pit at Crockerton worked the **Gault** for bricks; it is now flooded or built over. A pit at Donhead St Andrew, sited on Gault, was probably worked for brick clay.

Building stone, wall stones and tiles

Many limestones have been exploited. With the renewed interest in natural stone for building, it is important to know the former pit locations. They are described in ascending stratigraphical sequence: **Blue Lias** at Stoney Stratton just west of the district; **Spargrove Limestone** east of Stoney Stratton and south of Chesterblade, and **Junction Bed** at Westcombe, Batcombe, Milton Clevedon and Chesterblade.

The **Inferior Oolite** has been widely quarried. It is still worked in Doulting Quarries (Plate 1) just north of the district and Horsecombe Bottom, Hadspen (Plate 2). Old quarries are found at Chesterblade, Small Down Knoll, Lodge Farm, Horsehill Farm, Creech Hill, near Batcombe, north of Bruton, Lusty, Sunny Hill, near Compton Pauncefoot and Maperton. Extensions to Hadspen House were carried out with stone from Horsecombe Bottom [6547 3144] and Limekiln [6540 3150] quarries. Other large buildings include Sunny-Hill Girls' School, Cole (Richardson, 1916).

Fuller's Earth Rock was quarried near Higher Alham, Southill House, Henley Grove, Batcombe, Shepton Mont-

Plate 1 Yellow, massively bedded, biosparite of the Doulting Stone in Doulting Quarry. Hammer length 30 cm. (A15555)

Plate 2 Hadspen Stone in Hadspen Quarry. Nodular, biosparitic limestones of the Upper Inferior Oolite. Hammer length 30 cm. (A15540)

ague and Maperton. **Forest Marble** limestones were worked north-east of Higher Alham, south of Wanstrow, near Studley Farm, Witham Friary, Upton Noble (where many houses are built out of the limestone), near Henley Grove, north-east of Bruton, between Bruton and Redlynch, North Cheriton, near Bratton Seymour and Stoney Stoke.

Cornbrash was worked near Wanstrow, north-east of Witham Friary, west of South Brewham, at Upton Noble (out of which Bellerica Farmhouse [7216 3919] is built), in and around Wincanton, and around Stoney Stoke.

Cucklington Oolite is particularly suited to walling as it is a flaggy shelly limestone. It was worked near Stoke Trister, Cucklington, Quarr Cross and Kington Magna.

The lower part of the **Clavellata Beds** has been used extensively for building, walling and roadstone. It was recently worked in a large pit near Silton. Other quarries were at Langham, Silton, Bourton and near Zeals. The **Eccliffe Member** was dug for building or walling stone near East Stour, Eccliffe, Wyke, Milton on Stour and Silton.

The **Tisbury Member** has been extensively quarried for building stone in and around Tisbury. At Bridzor Farm [9310 2735], a quarry was reopened briefly to supply stone for the repair of Old Wardour Castle. The last working quarry, Tisbury Quarry [931 291] (Manners, 1971), closed in 1977 (Plates 3 and 4).

The Upper Greensand **Ragstone** was widely worked in the Horningsham–Longbridge Deverill area, Fonthill

Plate 3 Tisbury Member consisting of well-bedded, bioclastic limestone. Tisbury Quarry. Hammer length 30 cm. (A15547)

Plate 4 Tisbury Quarry (closed 1977) in well-bedded, bioclastic limestones of the Tisbury Member of the Portland Stone Formation. (A15533)

Gifford, east of Penselwood, near White Cross, east of Zeals, West Knoyle–Upton–Milton area and Shaftesbury. At Zeals, the bed was worked in at least one mine [7887 3140], which was driven an unknown distance westwards. **Melbury Sandstone** was dug for rough stone at Maiden Bradley, Norton Ferris, Search Farm and Mere. A cottage [9394 3912] on Corton Down is built out of **Chalk Rock**.

Agricultural lime and marl

In the Charnage Down chalk pit [837 330], near Mere, the **New Pit Chalk** and the **Lewes Nodular Chalk** are worked as uncalcined agricultural lime. Many other pits were dug in the Chalk for lime, particularly the **West Melbury Marly Chalk**, **Zig Zag Chalk**, **New Pit Chalk** and **Seaford Chalk**.

Many of the pits listed under 'Building stones' may also have been worked for lime. Kilns near Wanstrow, Studley and Bruton are evidence that the **Forest Marble** was burnt for lime. There was a kiln in a quarry [6925 2498] in **Lower Cornbrash** at South Cheriton.

Gravel

River terrace deposits border the River Cale, south of Wincanton, and along the River Wylye, but are generally very thin (less than 0.5 m thick), of poor quality and have over 1 m of overburden; nevertheless, they have been worked in small pits [e.g. 9400 4148 south-east of Knook].

At the western end of Great Ridge, clay-with-flints, particularly the basal beds, is sufficiently flinty to have been worked for gravel.

Roadstone

Inferior Oolite was dug [3565 3165] for stone for the base of runways at RNAS Yeovilton. Loose debris of the **Boyne Hollow Chert** (Upper Greensand) was probably dug as roadstone, such as from pits between Maiden Bradley and Sutton Veny, at Dead Maid Quarry, Mere, north-east of Pen Ridge Farm, near Upton and East Knoyle, Wincombe Park, on Semley Hill and near Donhead St Andrew.

Chalkstones of the **Lewes Nodular Chalk**, particularly the **Chalk Rock,** have been widely used for road base material and farm tracks: on Whiten Hill, between Brixton Deverill and East Knoyle, south of Tytherington, on Cotley Hill, north-east of West Hill, north of Knook and on Knook Horse Hill, near Hindon and White Sheet Hill.

Some of the very flinty **Clay-with-flints** [around 880 379] may have been worked for roadstone.

Building sand

Most sands in the district are too fine grained for commercial use. There were, however, some small pits: **Bridport Sands** near Yarlington; the basal **Hazelbury Bryan Formation** on Lawrence Hill; **Sandsfoot Grit** near Bourton; **Lower Greensand** south-east of Fonthill Gifford and north-west of Old Wardour Castle. Most sand came from the **Upper Greensand**, chiefly the **Shaftesbury Sandstone**, although some **Cann Sand** was also probably worked.

Millstones and Scythe stones

Around Penselwood, numerous conical-shaped pits ('Pen Pits)', 2 to 3 m deep, occur in two areas, now wooded [767 320, 766 315]. A few also occur at Zeals Row [773 318]. All are situated on **Shaftesbury Sandstone**. They were once more widespread; in the 19th century, over 700 acres were covered by 20 000 pits (Bates, 1905). It is thought that the pits were excavated for millstones and quernstones (Winwood, 1885). Winwood also refers to 'Penstones' as being blocks from the Boyne Hollow Chert which could be fashioned into scythe stones. Farmer (1992) records that mills at Taunton, Downton and central southern England obtained stones from Penselwood from at least the year 1208.

WATER RESOURCES

Surface drainage of the district is approximately radial (Figure 2) effected by the rivers Wylye, Frome, Brue, Cam, Yeo, Cale, Stour, Lodden and Nadder and/or their tributaries. Winterbournes are common on the chalk. These valleys are dry for much of the year and sustain streams only when the water table rises to coincide with the valley floor after winter recharge.

The average annual rainfall ranges from 875 mm in the south to over 950 mm on the high ground in the north-east. Potential evapotranspiration is about 525 mm, although actual evaporation is commonly less, since summer rainfall is generally inadequate to match high potential evaporation rates (Avon and Dorset River Authority, 1970).

The **Upper Greensand** and **Chalk Group** constitute the main aquifer, with the Gault acting as a basal aquiclude. Small quantities of groundwater occur in Jurassic sandstones and limestones, the intervening clays being aquicludes.

Licensed groundwater abstraction data, together with the uses to which the water is put are listed in Table 2. Chalk and Upper Greensand together provide over 96 per cent of the groundwater volume licensed for abstraction. The largest use is for public supply (over 50 per cent), the Cretaceous aquifer providing almost 98 per cent of the total. Many more sources, particularly springs, were formerly utilised for public supply to rural communities. Other major uses are watercress at Hill Deverill [868 405] and for fish farming (over 32 per cent and about 8 per cent of total abstraction respectively). Over 80 per cent of abstraction from Cretaceous strata originates from the Lower Chalk/Upper Greensand, with a further 17 per cent from a single Lower/Middle Chalk public supply source. Only very limited supplies are obtained just from the Lower Chalk. In addition to the quantities listed in Table 2, abstraction up to 1500 megalitres per annum (Ml/a) is licensed at Kingston Deverill [841 372] for river augmentation. The Upper Greensand also supplies some private systems, and appreciable quantities of water are obtained from Jurassic aquifers.

The volume of groundwater licensed for abstraction for agricultural use is small (5 per cent). About 68 per cent of this demand is met from Jurassic aquifers, mostly from springs (for example all the Portland Group sources); only small quantities are used for spray irrigation. The Bridport Sands is the most important Jurassic aquifer (almost 70 per cent of the Jurassic licensed quantities), and the springs at Pitcombe [673 329] are the only non-Cretaceous public supply source.

Jurassic strata

The more important aquifers are the Bridport Sands, Inferior Oolite, Fuller's Earth Rock, limestones in the Forest Marble, and Cornbrash, the Corallian and Portland groups. The aquifers are limited in extent and thickness and are truncated by major faults. Springs are common at faulted junctions and along the base of the aquifers. Although important as an agricultural water source in the west of the district, resources from Jurassic strata are limited due to constraints on storage and recharge. Significant decline in yield and water level may occur in response to pumping.

The **Bridport Sands** appear to be in hydraulic continuity with the overlying **Inferior Oolite** where confined beneath the Fuller's Earth. At outcrop, the Inferior Oolite mainly occupies elevated positions and drains into the Bridport Sands. Boreholes of 150 to 200 mm diameter provide yields

Table 2 Licensed groundwater abstraction data for the district.

| USE | PUBLIC SUPPLY | | PRIVATE SUPPLY | | AGRICULTURE | | | | | | FISH FARMING | | INDUSTRIAL COOLING | | TOTAL | |
| | | | | | GENERAL | | SPRAY IRRIGATION | | WATERCRESS | | | | | | | |
AQUIFER	Ml/a	Licences	Ml/a	Licences	Ml/a	Licences	Ml/a	Licences	Ml/a	Licences	Ml/a	Licences	Ml/a	Licences	Ml/a	Licences
Upper/Middle Chalk	2550.0	1	—	—	34.5	6	—	—	—	—	—	—	—	—	2584.5	7
Lower Chalk	—	—	1.7	1	64.3	11	—	—	—	—	—	—	—	—	66.0	12
Lower Chalk/Upper Greensand	3300.0	1	—	—	21.3	4	—	—	—	—	—	—	—	—	3321.3	5
Upper Greensand	2387.4	3	95.5	8	63.3	12	18.2	1	5053.0	2	1317.0	3	113.7	1	9048.2	30
Portland Group	—	—	—	—	106.5	5	—	—	—	—	—	—	—	—	106.5	5
Corallian Group	—	—	—	—	17.0	8	—	—	—	—	—	—	—	—	17.0	8
Great Oolite Group	—	—	3.8	1	62.8	19	—	—	—	—	—	—	—	—	66.6	20
Bridport + Pennard Sands	181.8	1	35.6	3	195.0	20	4.1	2	—	—	—	—	—	—	416.5	26
Totals	8419.2	6	136.6	13	564.7	85	22.3	3	5053.0	2	1317.0	3	113.7	1	15 626.6	113

Data supplied by the Environment Agency

of up to 1 l/s. Yields from springs also generally range up to 1 l/s, and rarely exceed 3 l/s. Boreholes are commonly constructed using plain surface casing with open hole below the water table, but there is no record of silting, possibly due to low pumping rates. It may, however, be necessary to use a screen and gravel pack to prevent sand ingress, particularly if higher pump rates are required. Few boreholes penetrate only the Inferior Oolite aquifer, but are drilled through into the Bridport Sands. Yields also range up to 1 l/s from small-diameter boreholes. Explosives have been tried to develop a borehole, but they did not improve the yield.

Small supplies of water are obtained from the **Fuller's Earth Rock**, impersistent limestones in the **Forest Marble**, and the **Cornbrash**. The Fuller's Earth, Frome Clay and clays of the Forest Marble are aquicludes. Boreholes, which commonly penetrate more than one of the aquifers, are mostly small diameter (up to 200 mm) and commonly cased throughout, with screen or perforated sections adjacent to aquifers. Yields from springs, boreholes and large-diameter wells rarely exceed 1 l/s (none exceed 2 l/s) and some boreholes have been abandoned because of poor yields.

The clays of the Kellaways and Oxford Clay formations constitute a major aquiclude beneath the Corallian Group aquifer. Locally, small quantities of groundwater have been obtained from shallow wells penetrating the weathered upper sections of these strata, but water quality is likely to be poor and is often ferruginous.

The majority of groundwater sources penetrating the **Corallian Group** are shallow, large-diameter wells, many of which have been used for local supply (Whitaker and Edwards, 1926). Yields are invariably poor, and rarely exceed 0.6 l/s, possibly due to restricted saturated thicknesses. Higher yields (4 to 12 l/s) were obtained from three 200 mm-diameter boreholes near Gillingham [8087 2612, 8086 2615, 8115 2613], where the Corallian aquifer is confined beneath Kimmeridge Clay. The latter is cased out in all three boreholes, with screen installed adjacent to the aquifer in two of the boreholes.

The yields of springs issuing from the **Portland Group** are highly variable, but range up to almost 8 l/s. Only one exploratory borehole has penetrated the Portland Group and, although reported to yield 1.3 l/s, the water was 'milky' and considered unfit for consumption.

Water quality information for the Jurassic aquifers is sparse, mostly being described simply as 'good' or 'pure'. Analyses suggest generally hard groundwaters (up to 300 mg/l as $CaCO_3$), but from some confined limestone units the water could be very hard (> 1000 mg/l as $CaCO_3$). Chloride ion concentrations are generally less than 20 mg/l. At outcrop, the fissured Jurassic aquifers are highly vulnerable to contamination.

Cretaceous strata

The **Upper Greensand** and **Chalk Group** together constitute the main aquifer in the district, but the Upper Greensand is confined beneath the West Melbury Marly Chalk. At Brixton Deverill [863 388], artesian conditions were encountered on penetrating the top of the Upper Greensand, but the confined piezometric level was only 0.2 m above that of the Lower Chalk water table. During the testing of a nearby production borehole, water levels equalised, and borehole logging indicated some degree of hydraulic continuity, with groundwater moving downwards from the fissured base of the Lower Chalk into the Upper Greensand. Although the Upper Greensand is generally confined beneath the West Melbury Marly Chalk, it is not easy to distinguish yields from each formations in boreholes that penetrate both formations. They are therefore considered as a single aquifer.

The Cretaceous aquifer water table forms a subdued version of the surface topography. The highest water-table elevations, above 180 m OD, are along the north-west margin of the Cretaceous outcrop (Figure 3) and decline eastwards to about 100 m OD in the south-east and to about 90 m OD in the north-east (Institute of Geological Sciences, 1979).

Rest water levels have been regularly monitored in several boreholes. The longest, fairly continuous, records are for Fonthill Bishop (Boreholes OB1 [941 342] and OB2 [9471 3417]). These show regular seasonal water-level fluctuations in response to recharge (Figure 4), the largest fluctuations occur in boreholes located on higher ground, borehole OB1 being at the higher elevation. As there is little surface run off, infiltration, which recharges the saturated zone below the water table, is approximately equal to rainfall less evapotranspiration. Recharge therefore occurs almost entirely during the winter months, when rainfall is normally at a maximum and evapotranspiration at a minimum.

The degree of cementation of the Upper Greensand strata is variable. Intergranular groundwater flow occurs in poorly cemented sands, whilst water storage and movement, commonly massive sandstone units; permeability is dependent on the number, size and degree of interconnection of fractures. The Boyne Hollow Chert, for example, can be highly permeable where it is fractured, rapidly draining leakage from the overlying, less permeable, Lower Chalk. Where not fractured, it acts as a hydraulic barrier and springs may issue from near its top. Boreholes penetrating Upper Greensand require properly designed sand screen and gravel pack construction to limit sand ingress.

Yields from the Upper Greensand, from both narrow (up to 200 mm) and large-diameter boreholes and wells, are generally less than 1 l/s, commonly from a limited saturated thickness. Higher yields have been obtained from some wells of over 1.5 m diameter, with 4 l/s being obtained at Wolverton [7880 3164].

Many springs issue from the Upper Greensand and, whilst many yield less than 1 l/s, yields range up to almost 10 l/s. The larger springs, many of which were used for local public supplies (Whitaker and Edmunds, 1925), continue to flow even during prolonged dry periods, although yields may decline by between a third to a half. The largest recorded yields, between 10 and 24 l/s, are from three overflowing shallow wells (up to 2.5 m deep), probably enhanced springs, near Motcombe [869 256].

The Chalk is a microporous limestone in which water flow is mainly along fissures and joints. Matrix porosity is commonly in the range of 25 to 45 per cent, but makes a minimal contribution to total groundwater flow. Many fissures and joints cut the Chalk, but most water flow occurs along a few discrete horizons or 'secondary fissures' which result from enlargement of joints (usually bedding-plane joints) by solution and weathering (Foster

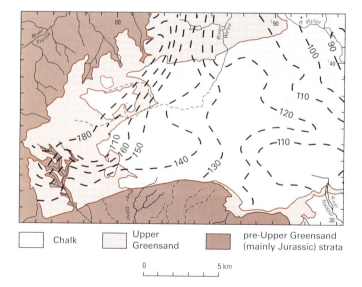

Chalk   Upper Greensand   pre-Upper Greensand (mainly Jurassic) strata

0   5 km

Figure 3 Potentiometric surface of the Chalk and Upper Greensand aquifer. Contours in metres relative to Ordnance Datum (adapted from IGS, 1979).

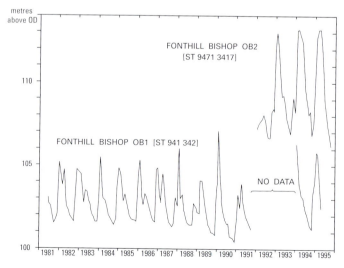

Figure 4 Hydrographs for the period 1981 to 1995 for two observation boreholes at Fonthill Bishop.

Table 3 Typical chemical analyses of groundwaters in the district.

Location NGR	Mere Waterworks* ST 8206 3262	Hill Deverill† ST 8680 4030	Donhead St Mary* about ST 905 245	Shaftesbury* ST 8647 2320
Type of source	Borehole	Borehole	Spring	Well
Aquifer	Lower Chalk/ Upper Greensand	Lower Chalk/ Upper Greensand	Upper Greensand	Upper Greensand
Date of analysis	28.4 1948	1.2.95	about 1911?	21.1.1919
pH	7.2	7.4	NR	NR
Conductivity (μS)	430	497	NR	NR
Total dissolved solids mg/l	282	NR	120	280
Calcium (Ca^{2+}) mg/l	89	106	55	82
Magnesium (Mg^{2+}) mg/l	2.4	1.8	0.7	1
Sodium (Na^+) mg/l	6	NR	14.5	18
Potassium (K^+) mg/l	NR	1.1	NR	NR
Carbonate (CO_3^{2-}) mg/l	126	163	70	110
Sulphate (SO_4^{2-}) mg/l	13	16	24	25.5
Chloride (Cl^-) mg/l	10	13	17	17
Nitrate (NO_3^-) mg/l	8	4.9	9	18.5
Silica (SiO_2) mg/l	26	24	10	8

Data sources:

* National Well Record Archive

† National Rivers Authority, Wessex Region

NR: not reported

and Milton, 1974; Price et al., 1982). It is generally considered that fissure density is greatest at lower elevations due to stress release due to overburden removal. Consequent solution enlargement of fissures and relatively shallow water tables provides the best hydrogeological conditions for high-yielding boreholes. The hydraulic properties of the Chalk also vary according to the marl content, with the Upper Chalk generally providing higher yields than the more marly Middle and Lower Chalk, especially the West Melbury Marly Chalk.

Yields from just the Lower Chalk range from 0.3 to 2 l/s, although the majority of wells and boreholes penetrate a saturated thickness of less than 10 m. Boreholes with similar yields from greater saturated thickness are commonly sited on higher ground outside the valleys. The highest yields were obtained from a pair of wells at Burton Field, Mere [821 327]. A large-diameter (1.8 m) well penetrated a saturated aquifer thickness of about 7.5 m, yielding over 13 l/s for a drawdown of about 5.6 m. A second well (1.5 m diameter), in the base of which a 300 mm borehole was drilled, penetrated a saturated thickness of 18 m and yielded over 25 l/s for a drawdown of only 4 m. A nearby, 1.8 m-diameter well yielded over 12 l/s from a saturated thickness of only 4.5 m.

Boreholes penetrating both the saturated Lower Chalk and Upper Greensand almost invariably have higher yields than those penetrating only one of the formations. Those on higher ground, away from valleys, yield between 1 and 4 l/s, whilst boreholes on the flanks of the Wylye valley south of Heytesbury yield between 5 and 6.5 l/s. Yields from large-diameter boreholes in valleys are, however, considerably higher. At Brixton Deverill [857 380], boreholes ranging in diameter from 520 to 760 mm provided yields between 100 and 190 l/s for drawdowns of up to 31 m. At Burton Field, Mere, a 910 mm diameter public supply borehole [821 326] yielded 126 l/s for a drawdown of about 12 m (as compared to 25 l/s from a nearby large-diameter well penetrating only the Lower Chalk).

Boreholes penetrating both Middle and Lower Chalk or only the Middle Chalk are sited mostly on interfluves and yields rarely exceed 2 l/s. At a lower elevation, near Fonthill Bishop [941 342], two public supply boreholes of over 650 mm diameter which penetrate all three Chalk formations, although only the basal few metres of the Upper Chalk is saturated, yielded 88 and 85 l/s with drawdowns of 14.5 and 5 m respectively. A long-term aquifer test to investigate the effect of continuous abstraction on stream flow determined that about 70 per cent of abstraction was derived from groundwater storage and that environmental impacts would be negligible. The base of the Upper Chalk lies above the water table over much of the district and even where its base is saturated, no borehole penetrates this without continuing into the underlying Chalk.

Few properly conducted aquifer tests have been carried out on boreholes in the district and, consequently, transmissivity and storage coefficient values for Cretaceous strata are few. An aquifer test at Brixton Deverill [863 388] gave a transmissivity value of 8200 m^2/d and storage coefficient of 1.3×10^{-2} for the combined Upper Greensand/ Lower Chalk aquifer, assuming that the two formations act as a single aquifer. An alternative analysis gave separate transmissivity and storage values for the Upper Greensand and Lower Chalk of 260 m^2/d and 1.7×10^{-4} and 8790 m^2/d and 9.9×10^{-3} respectively (Avon and Dorset River Authority, 1973). Tests on three boreholes north of Heytesbury [about 927 446] (just outside the district), gave average transmissivity and storage values for the Upper Greensand and Lower Chalk of 285 m^2/d and 1.8×10^{-4} and 70 m^2/d and 7.0×10^{-5} respectively. Chalk aquifer properties on Salisbury Plain indicates that transmissivity values range from 50 to 8200 m^2/d, with a geometric mean and median of 1400 and 1600 m^2/d respectively. Storage coefficients range from 1×10^{-4} to 0.05 (MacDonald and Coleby, in press).

Full major ion analyses of Cretaceous groundwaters in the district are sparse. Representative analyses, two from the Upper Greensand and two from the combined Upper Greensand and Lower Chalk are shown in Table 3. Groundwater is generally potable and of calcium bicarbonate type. Water from the Chalk is invariably hard, with total hardness ranging up to about 300 mg/l (as $CaCO_3$). Water from the Upper Greensand is less hard (total hardness commonly less than 100 mg/l), particularly in recharge zones remote from the Chalk outcrop. Chloride ion concentrations are generally less than 20 mg/l. The few determinations of nitrate concentrations in the district do not generally exceed 10 mg/l. In common with other parts of England underlain by Chalk strata on which nitrogen fertilisers are applied, leaching of nitrate to groundwater is possible.

LANDFILL AND WASTE DISPOSAL

Some 27 landfill sites have operated in the district, of which only two remain open. Most are located on the clay formations, notably the Gault, Kimmeridge Clay, Oxford Clay and Forest Marble, and accepted mainly inert non-hazardous and household wastes. One site on Chalk (Sheep Wash Lane, East Knoyle [887 311]), accepted only household waste and closed in 1987. The Wistley Farm (Silton Quarries) site [780 285], located on Corallian Group strata, is licensed to accept non-hazardous construction waste and small quantities of garden waste. The second operational site, near Maiden Bradley [798 389] on Upper Greensand (Boyne Hollow Chert), accepted domestic waste in the late 1950s and, since 1980, has been licensed for the disposal of small amounts of construction and farm wastes.

FOUNDATION CONDITIONS

The suitability of bedrock and superficial deposits for the foundation of buildings and structures is dependent on their geotechnical properties and their mass geological characteristics. The rock and engineering soil types of the area cover a range of geotechnical materials including mudrock, sandstone and limestone, and their weathered products. The following comments are intended as a general guide to their likely behaviour, but may not take into account local variations in lithology or behaviour. They are not intended as a substitute for an appropriate site investigation. The account is based largely on lithological descriptions of the rock units without a detailed analysis of the limited geotechnical data available in site-investigation reports.

Unless otherwise stated, where unweathered, the following list of rock formations should offer good foundation conditions with good bearing capacity and low settlement: Spargrove Limestone, Bridport Sands, Junction Bed, Inferior Oolite, Fuller's Earth Rock, limestones in the

Forest Marble, Cornbrash, Cucklington Oolite, Sturminster Pisolite, Clavellata Beds, Upper Greensand (except the Cann Sand). However, with relatively thin deposits, foundation conditions will also depend on the underlying beds. Excavation may require ripping and the use of hydraulic or pneumatic rock breakers, depending on the degree of cementation, fracturing and state of weathering of the material.

Detailed lithological descriptions of the following units are given in the succeeding chapters.

Blue Lias consists of an alternating sequence of limestones with thin mudstones. Locally, beds of mudstone, up to 1 m thick, occur and may weather to high plasticity clay; if thick beds are present, some settlement will occur. Foundations may need to be designed to take account of the shrinking and swelling of the clays due to changes in water content.

Pylle Clay consists of up to 20 m of dark grey shaly mudstone. *Ditcheat Clay* and *Down Cliff Clay* consist of pale to medium grey, silty mudstone and silt. *Pennard Sands* consist of silty and fine-grained sandy clay, clayey, micaceous, fine-grained sand and thin beds of sand. There are no geotechnical data for these members. Small slips occur on the Pylle and Ditcheat clays and Pennard Sands on Pennard Hill west of the district. Excavation may be achieved by digging, but excavation stability is probably poor.

Lower Fuller's Earth consists of olive-grey, silty, predominantly fissile, weakly calcareous, shelly mudstone. There are no geotechnical data from the district, but no slip has been recorded on the member. In the Shaftesbury district, the clay minerals consist dominantly of kaolinite and illite; smectite is present in the lower part of the member, and chlorite towards the top (Barton et al., 1993). Where unweathered in flat-lying areas, the member should offer good foundation conditions with good bearing capacity. Excavations may be accomplished by digging.

Frome Clay consists mainly of buff and olive-grey calcareous clays. The clay minerals consist of illite and kaolinite, with only minor amounts of chlorite. Shallow, possibly translational, degraded slips occur on the steep slopes of Frome Clay below the Forest Marble. Excavation may be achieved by digging, but because of the possibility of degraded slip material at the surface, care needs to be exercised to ensure that bedrock is reached.

Forest Marble comprises lenticular limestone, sand and sandstone in a dominantly mudstone sequence. The ratio of the lenticular beds to clay varies greatly from 1 mm-thick limestones, interbedded with 2 to 3 mm clay, to clay-free limestone up to 11 m thick. Clay minerals are mainly kaolinite and illite. Samples range from low to high plasticity, but most are of intermediate plasticity. The mean undrained shear strength of samples from the Wincanton By-pass is 73 kPa. Swallow holes have been recorded on the outcrop west of South Cheriton [around 6765 2480].

Kellaways Formation consists of clayey sand and sandstone, and sandy clay and mudstone. The formation is not always distinguished from the Oxford Clay in site-investigation boreholes. Excavation may be achieved by digging, but stability is probably poor.

Oxford Clay comprises grey, calcareous mudstone, overlying 5 to 15 m of bituminous mudstones (Peterborough Member). Locally, there is a basal lenticular unit of grey, calcareous, silty, shelly mudstone, up to 10 m thick (Mohuns Park Member). The formation is not always distinguished from the Kellaways Formation in site-investigation boreholes and so the engineering parameters attributed to both units are compound. The dominant clay mineral is thought to be illite; mudstones are of intermediate to high plasticity. Landslips are widespread on the scarp face below Corallian beds. The undrained shear strength of samples from the Wincanton By-pass range from 57 to 323 kPa, with a mean (of 31 samples) of 118 kPa).

Hazelbury Bryan Formation and *Newton Clay* consist mainly of stiff and very stiff, sandy clay, with some interbedded very fine- to medium-grained sand which give rise to pronounced springlines [e.g. around 7656 2380]. In the Shaftesbury district, smectite is the dominant clay mineral; kaolinite is also common, mainly in the sandier parts of the sequence (Bristow et al., 1995). The clays vary from low to high plasticity, but are mostly of intermediate plasticity; they are not thought to be subject to frost heave. Large slips associated with the Oxford Clay occur south of the Mere Fault. Small, active slips occur north-west of Maiden Bradley. Excavation may be achieved by digging, but stability is probably poor.

Sandsfoot Formation is locally divisible into two members. The lower, Sandsfoot Clay consists of very stiff, fissured, shelly, locally sandy and ferruginous, silty clay and is dominantly of intermediate plasticity. The upper, Sandsfoot Grit, consists of silty, fine-grained sand and fine-grained sandy clay and may be frost susceptible. Both members may be excavated by digging, although some ripping may be required where thin sandstone and ironstone beds are present. Some running sand may occur.

Kimmeridge Clay consists of grey, stiff and very stiff, carbonaceous, fissured, partly calcareous and pyritic and, in places, bituminous mudstone; cementstone beds and septarian nodules occur. The mudstone weathers to a firm to stiff, mottled grey and brown, heavy, poorly drained clay. Weathering extends to between 2 and 4 m. Gypsum, as well-formed crystals up to 100 mm long, is locally common in the weathered zone and may result in volume changes. Clay mineralogy consists of about 50 per cent illite and equal amounts of kaolinite and 'expansible' minerals [smectite] (Gallois, 1979). Clays are mostly of high plasticity. Slips, some active, occur on some steep slopes. The formation can be excavated by digging, but stability will be poor in the weathered zone.

Portland Group comprises the Wardour Formation overlain by the Portland Stone Formation. The base of The Wardour Formation is commonly marked by springs. Excavation should be possible by digging, but stability could be a problem, especially in the water-saturated basal sand. The Portland Stone consists variably of micrite, fine-grained, calcareous sandstone and sandy limestone and oolite; chert nodules vary from sparse to common. Widespread cambering of the limestone over the Wardour Formation occurs. Jointing is common, with some joints up to 10 cm wide. Development on valley sides should take account of the valleyward dip of the strata due to cambering and the higher incidence of joints. Sink holes are developed on the Tisbury Member south of Ruddlemoor Farm, Hindon [9060 3046 to 9076 3028], and on Gault where it overlies the Tisbury Member near Fonthill Gifford [9211 3135, 9263 3109, around 9285 3070].

Lower Greensand consists of glauconitic, silty, fine- to medium-grained, locally clayey sand; pebbly sand and ferruginous sandstone occur in places. The formation base is commonly marked by springs [e.g. 9041 2620 to 9074 2642, 9091 2639 to 9088 2627]. Excavation is possible by digger, but stability is a problem, particularly where the sands are water-saturated.

Gault consists dominantly of firm to stiff, grey, glauconitic, micaceous, very silty, fine-grained sandy clay. In places, as on Round Hill [915 260], there is a thin (up to 5 m) bed of clayey, glauconitic, fine-grained sand. The base is locally marked by springs. Smectite makes up 20 to 40 per cent of clay minerals (Bristow et al., 1995, fig. 45). The clays are of intermediate plasticity. Landslips are common and are spectacular with morphological features indicative of historically recent activity. The landslipped slopes are close to limiting equilibrium; reactivation could easily be caused (see 'Landslips and ground stability' below).

Upper Greensand sands and sandstones have a wide range of strengths from massive, well cemented and very strong (Ragstone), to moderately strong (parts of the Shaftesbury Sandstone, Boyne Hollow Chert and Melbury Sandstone, depending on cementation), to weakly cemented or uncemented (Cann Sand). The outcrop of the **Cann Sand** generally forms a shelf-like, well-drained platform, commonly with springs at the base. It is extensively landslipped (Plate 5) (see 'Landslips and ground stability'). In places (Figure 5), the Cann Sand is cambered over Gault. **Shaftesbury Sandstone** consists of alternating beds of fine-grained, glauconitic sand and weakly calcite-cemented sandstone in the lower part, and hard, shelly, calcite-cemented, glauconitic sandstone (Ragstone) in the upper part. The Ragstone is commonly jointed, with some joints being open. Some joints may result from cambering; others, such as the water-bearing ones at Wolverton (Freshney, 1995b) are related to the Mere Fault. Sandstones in the **Boyne Hollow Chert** weather to give a rubbly mixture of chert and siliceous sandstone in a sandy and, commonly, clayey matrix. **Melbury Sandstone** consists of fine-grained, fossiliferous, calcareous sands and sandstone. The sulphate content of the Cann Sand is low (Class 1); that of the Shaftesbury Sandstone and Boyne Hollow Chert is probably similar.

Chalk Group comprises a variable suite of lithologies. Soils on the **West Melbury Marly Chalk** are poorly drained. The marls are of low plasticity. The Chalk Rock near the base of the **Lewes Nodular Chalk** consists of about 1 m of dense, hard, porcellanous chalk with glauconitised hardgrounds; the rock is usually cut by widely spaced joints. All the members above the West Melbury Marly Chalk should offer good foundation conditions with good bearing capacity. Excavation should be by digger, except in the Chalk Rock which may require ripping. Excavations should be fairly stable. Many chalk pits still retain vertical

Plate 5 Landslips in Gault and Upper Greensand. In the foreground is slipped Gault and Upper Greensand, but it is only the fields in the middle distance which are characteristically hummocky. Thanes Farm, Motcombe. (A15127)

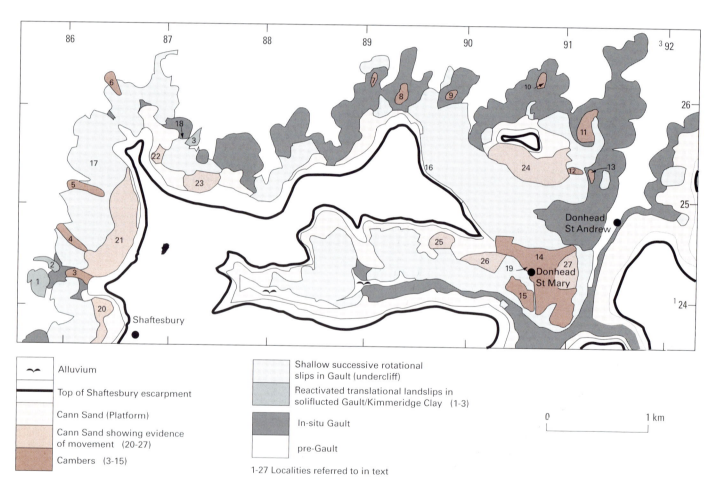

Figure 5 Slope morphology and mass movements in the district.

faces long after abandonment, although the chalk at the top of the pits tends to be rubbly. Sink holes occur in places on the Seaford Chalk.

No geotechnical data are available for the superficial deposits. The geotechnical properties of *Head* deposits are dependent on those of the parent material, but, in general, the deposits are in a remoulded or loose condition, weak and compressible. *Alluvium* commonly has a firm to stiff, desiccated crust; the clays vary from intermediate to very high plasticity. Excavation in most drift deposits can be achieved by digging, but support will probably be needed.

LANDSLIPS AND GROUND STABILITY

The largest and most extensive landslips in the district are associated with the Gault and Upper Greensand, and Oxford Clay and Hazelbury Bryan Formation. Smaller slips occur on the Frome Clay, Forest Marble, Kellaways Formation and Kimmeridge Clay.

Many slips in the **Frome Clay** occur on the steep scarp face below the Forest Marble. Most are degraded and probably predate the last phase of head formation. They were produced by sliding of Forest Marble limestone and sandstone debris on the underlying clays. Flowstone and solution cavities in limestones in a quarry [6695 2420], just west of Charlton Hill, indicate a higher water table in the past. The largest slip, about 1 km long, occurs north-east of Higher Alham [681 423 to 6845 4145].

Landslips are widespread in the **Oxford Clay** and **Hazelbury Bryan Formation** on the steep face below the Stour Formation. Slips were probably initially of stepped or successive rotational type, but pass downslope into translational ones. Between Buckhorn Weston [7600 2460] and Stoke Trister [740 290], slips are related to springs issuing from sands in the Hazelbury Bryan Formation. In places, there are well-defined back scars (for example on Coneygore Hill, Stoke Trister, and north of Buckhorn Weston [7571 2595 to 7584 2553]) and prominent toes (e.g. between Clapton Farm and Buckhorn Weston). Despite their fresh morphology, there are few signs of recent movement; an exception is the fresh back scar on Coneygore Hill. Small, active, slips in Oxford Clay along the Stour Valley [e.g. 786 240, 766 250] and Filley Brook [764 263 to 765 248], and in the Hazelbury Bryan Formation in the valley north-west of Maiden Bradley [7950 4118 to 7975 4092, 7975 to 8022 4222] are probably caused by fluviatile undercutting of the foot of the slope, combined with saturation of the clay by spring water.

Slips on steep slopes of the **Kimmeridge Clay** south-west of Sedgehill [around 853 273] and south of West Knoyle [860 307 to 872 300] show signs of recent movement; most of these result from erosion by streams bounding the slips. Other slips relate to springs from the base of the Gault south of West Knoyle [863 325, 868 310, 873 305, 875 303], the last showing signs of recent movement, and south of Semley [8900 2635], and with springs from the base of the Portland Beds near West Tisbury [905 293 to 906 284].

Landslips on the **Gault and Upper Greensand** are spectacular, with back scarps, undulating irregular surfaces, ponds, and prominent toes. Some slips in the Shaftesbury area were mapped and divided into morphological types (Bristow et al., 1995, fig. 61); this mapping has been extended into the Wincanton district (Figure 5). The reader is referred to the above memoir for details of the development and evolution of these landslips.

Three distinct landforms can be recognised; they are the *Scarp, Platform, and Undercliff.*

Scarp, 20 to 30 m high (Plate 5), with average slope angles of between 30 and 35°, occurs continuously along the Upper Greensand outcrop. The scarp face, cut in Shaftesbury Sandstone and capped by Boyne Hollow Chert, is smooth, vegetated and divided into broad arcuate embayments up to 1.5 km across. There is no evidence of instability on the scarp, but some slips [7530 3145], south-west of Penselwood, cut back to the Shaftesbury Sandstone.

Platform, up to 250 m wide, slopes from the base of the escarpment at between 3° and 7°. The inner edge, at the Shaftesbury Sandstone base, forms a continuous negative feature break. The outer edge is irregular, but is usually marked by springs and a break of slope. The platform is developed on uncemented or poorly cemented Cann Sand. The topography varies from smooth, to gently undulating (Figure 5, localities 20–27), to hummocky, with each type grading into each other (Plate 5). Hummocky surfaces are more noticeable near the outer edge where the slope angle steepens slightly. The platform is generally well drained, commonly with springs at the outer margin. In places (Figure 5, localities 4–15), for example between embayments, smooth spurs or detached masses of Cann Sand extend down to the clay vale, suggesting that the Cann Sand is cambered over Gault.

Undercliff The edge of the platform coincides approximately with the Gault/Upper Greensand junction. In most places, the undercliff comprises an upper, steeper (9° to 12°) degrading zone, and a lower accumulation zone with lesser (7 to 10°) slope angles. Slides in the erosional zone are of shallow, successive rotational type. The depth of slipping is probably up to 6 m in places. Some disturbed slopes below the undercliff are flatter and more convex, for example below Cowherd Shute Farm [857 242] (locality 1), where the overall angle is about 6.5°. This is interpreted as a reactivated translational slip. In addition, east of Pottles Hill Plantation [858 244] (locality 2), there are translational slides on slopes of about 8.5°.

There are few signs of present-day movement, but fresh back scarps, tilted trees, and toe features indicate historically recent activity. The toes below [7675 3974] and south-west [7630 3922 to 7612 3910] of Witham Park Farm are over 2 m high. The local farmer observed that the former slip 'is on the move'. The latter, part of a 2 km-long series of slips [760 375 to 7690 3935], has a very fresh, highly irregular, surface and is clearly active. In places, small, temporary ponds occur between slips, for example in Coneygore Wood [749 322], north-west of Penselwood. Locally, for example Cockroad Wood [748 323], the slips form a series of elongate rotated blocks, in which Gault and Cann Sand are repeated down slope. In the hollows between, marshy areas or ponds occur. In the large slip below Donhead Cliff, there is a pronounced hollow [8953 2566–8957 2550] (locality 16) at the back of the slip at the escarpment foot. Recent movement has occurred near Kingsettle [around 8627 2551] (locality 17) and Hart Hill (locality 18) farms, where there is a fresh back scar [8708 2559–8720 2563].

Translational slipping of the Cann Sand on the Platform

Crescentic sand steps on the platform, about 1 to 2 m high and separated by 10 to 15 m of flat ground, so well developed around Shaftesbury (Bristow et al., 1995), have also been noted [e.g. 906 243] (locality 19) in this district. They are possibly the result of movement along low-angle shear surfaces that follow, or are close below, the Gault/

Upper Greensand junction. A similar mechanism explains comparable features in the Upper Greensand and Gault on the Isle of Wight (Hutchinson, 1981).

Future stability may depend on variations in water table and on changes in hydraulic properties at seepage faces, especially close to spring lines near the platform edge. Slipped slopes of the undercliff are close to limiting equilibrium. Reactivation could easily be caused by rising groundwater levels in the sand platform, by cutting into the slope or loading upper slopes with fill. Thorough site-investigations to determine stability of slopes are necessary prior to any construction.

Cambering

Cambering probably developed in perennially frozen ground in a periglacial climate; it is usually associated with overconsolidated clays with high lateral stresses. Valleyward flow of clay from under the cambers and its removal by stream erosion appears to be an essential part of the process. Small-scale faulting is usually developed in overlying strata.

The base of the Upper Greensand, and locally the Gault, on some smooth-surfaced spurs in landslipped ground falls steadily away from the scarp, for example near Stourton [762 338] and farther east [around 906 241, 8907 2625, 8935 2605, 8985 2608]. The base of the Cann Sand in the largest camber, at Donhead St Mary (Figure 5, locality 14), falls from 165 m OD, south-eastwards to 120 m OD over a distance of 1 km. In places, Cann Sand cuts across Gault to rest on Kimmeridge Clay, such as east of Bittles Green (locality 4), where Cann Sand descends from almost 175 to 135 m OD [8635 2506 to 8598 2519], and east of North Hayes Farm (locality 6) [8652 2608 to 8638 2631], where the base falls from 150 to 120 m OD. The base of a Gault outlier [7643 3922] below Witham Park, marked by a well-defined spring line, lies 20 m below its predicted level.

At outcrop, the Wardour Formation is much thinner compared to that in the Tisbury Borehole and is thought to be due to cambering of the Tisbury Member across the Wardour Formation. The valleyward dipping Tisbury Member was seen in a pit [9351 3107] at Fonthill Gifford, just south-east of Pythouse [9074 2847], in the road cutting [9118 2813] east of Hatch House, and on the north side of Oddford Brook [9412 2930].

Valley bulging

An exposure [7092 3567] south-west of South Brewham, reveals mudstones deformed by an asymmetrical north-east-verging box fold, with a steep short limb, a horizontal fold axis trending approximately north-west parallel to the valley, and an axial plane oriented 140/55° SW.

UNDERMINING, SOLUTION SUBSIDENCE AND COLLAPSE STRUCTURES

Shaftesbury Sandstone was worked for building stone in two adits at Wolverton [around 7887 3140]. Their underground extent is unknown.

Around Penselwood, there are numerous conical-shaped pits, 2 to 3 m deep known as 'Pen Pits' from which millstones and scythe stones were dug. They occur in two main areas, now wooded, north-east [767 320] and east of Penselwood [766 315]; a few also occur at Zeals Row [773 318]. Formerly more widespread, many have been destroyed by ploughing. Their underground extent is not known; they may have been bell pits.

Just east of the district, Portland Limestone has been extensively exploited underground. The gallery roofs are now unstable and there are high radon values in the galleries. There is only one record of underground working in this district, a cave [9366 3159] on the east side of Fonthill Bishop lake (Bristow, 1995b).

Shallow, circular, depressions which are the surface expression of swallow holes occur in the floor of some dry valleys on the Chalk, commonly in areas where clay-with-flints caps the interfluves [8707 3755, 8778 3853, 8835 3650, 8865 3655, 8864 3540, 8911 3457, 9102 3194, 9106 3202, 9014 3305], on the Tisbury Member southwest of Hindon [9060 3046 to 9076 3028], and on Gault where it overlies the Tisbury Member near Fonthill Gifford [9211 3135, 9263 3109, around 9285 3070].

MADE AND WORKED GROUND

There has been little large-scale mineral extraction in the district and therefore there is no extensive worked ground. There was, however, much small-scale activity. The more important of these pits and quarries are mentioned in the section dealing with Mineral Resources (see also Open-file reports listed in Information Sources (p.92)). Many have been backfilled, others partially filled and degraded, some are flooded.

There are some 27 waste disposal sites in the district which are either operational or have closed within the last 30 years. These are discussed in 'Water Resources' (p.9).

The Gillingham Gas and Coke Company [8059 2665] existed from 1837 until about 1980. Tar was produced as a by-product at this site and used locally to form a damp-proof course (Howe, 1983). Such industrial activity is known to produce ground contamination elsewhere, but any soil contamination will be partly buried beneath the inner ring road which crosses the site.

RADON

Radon is a naturally occurring radioactive gas produced by the decay of uranium and thorium which are present in all rocks. The gas can be a potential health hazard in buildings if high levels accumulate, and it has been estimated that at least half of the total radiation dose for the average Briton is obtained from combined radon and thoron (Clarke and Southwood, 1989). There is a direct link between geology and the radon levels generated at the surface.

Radon is easily dispersed to very low levels in the atmosphere. However, in confined spaces in contact with rock, such as caves, mines and buildings, radon and its daughters can accumulate. Soil gas is drawn into a building by a slight underpressure indoors which results from rising warmer air. Hence, radon problems in houses are due to bulk flow of ground gas carrying radon, with a relatively small contribution by the diffusion of radon through and from building materials. The amount of ground air drawn into a house will vary according to the local ground permeability and the nature of the leakage into the house.

In January 1990, the National Radiological Protection Board (NRPB) issued revised advice on radon in homes, resulting in a new Action Level for radon of 200 Becquerels per cubic metre (Bq/m^3). Areas in which there is one per cent or more of the homes above the Action Level should be regarded as 'Affected Areas', in which it is recommended that the government should designate localities where precautions against radon in new houses are required. A classification of areas requiring some form of building control has been adopted by the Building Research Establishment. If more than 10 per cent of existing houses are affected (high risk), all new dwellings in the area should incorporate full radon protection (i.e. a radon-proof barrier), along with provision for sub-floor ventilation measures if barriers prove ineffective. Where between 3 and 10 per cent of houses are affected (moderate risk), new dwellings should have provision for future sub-floor ventilation if required. In such cases, houses would be built complete with sump and extract pipe which could be connected to a fan if measurement revealed a problem with radon ingress. The low-risk class is equivalent to a zone in which more than 3 per cent of houses exceed the Action Level.

Areas with high levels of radon can be recognised: firstly, by measuring gas levels in houses: secondly, by estimating the radon-emission potential of the ground by measuring radon concentrations in soil gas, combined with geological and other factors that influence the surface emission of the gas, such as permeability and uranium concentration. In Somerset, there is generally good agreement between radon potential class (based on soil-gas radon and permeability), and house-radon risk class (based on measurements in houses) for most geological units (Appleton and Ball, 1995). The resulting Radon Potential Map is based on a classification of geological units using all soil-gas radon, permeability and house-radon data.

Geological units with High Radon Potential include the Bridport Sands, Inferior Oolite and Upper Greensand. Units with Moderate Radon Potential include Blue Lias, Corallian and Portland groups, Purbeck Limestone, clay-with-flints, river terrace deposits, alluvium, and head. The Spargrove Limestone, Cornbrash, Fuller's Earth Rock and limestones in the Forest Marble are probably too thin to present much of a problem since they are sandwiched in clays and can probably mostly be classified as Low Potential, with some isolated 'Highs', since they produce high levels of interstitial radon. The Lower Chalk probably shows a similar pattern. All other formations and members can probably be classified as Low Potential.

Geological radon-potential mapping cannot predict whether existing or new houses will be affected, since this depends on factors such as house use, ventilation and construction methods. However, it can delineate potentially affected areas accurately and should be more precise than using house-radon data averaged over 5 km squares.

Householders concerned about radon levels in their property are advised to contact the NRPB who will advise on whether measurement for levels of radon should be carried out. If measurement is advised, they will be carried out without charge by the NRPB (as at 1996).

CONSERVATION

Exposures of rocks and their fossils which demonstrate the geology and geomorphology, together with woodlands, meadows and wetlands, are an important resource for education, research and recreation. Sites can be protected by their designation as a Site of Special Scientific Interest (SSSI) or Regionally Important Geological/Geomorphological Site (RIGS), both administered by English Nature. Sites are notified to the Department of the Environment, local authorities, and site owners. There are five geological SSSIs in the district.

Shepton Montague railway cutting [686 316]. This cutting exposes one of the most important sections in the Fuller's Earth Rock in England. It is a key locality for the study of three Bathonian ammonite zones.

Godminster Lane Quarry and railway cutting (Lusty cutting of Richardson (1917)), Bruton [682 345]. This is an important locality for the Inferior Oolite, especially the thin remnants of Lower and Middle Inferior Oolite (*laeviuscula, discites* and *concavum* zones).

Bruton railway cutting [688 348]. This cutting is internationally recognised as the type section of the Middle Bathonian *morrisi* Zone, but equally important, sedimentary rocks of *subcontractus* Zone are well represented.

Dead Maid Quarry [803 324], Mere. This quarry provides the finest section in south-west England of the Upper Greensand/Chalk transition. It is a key locality for the study of the palaeontology of the West Melbury Marly Chalk, Melbury Sandstone and Boyne Hollow Chert.

Charnage Down chalk pit [837 329]. The pit provides superb exposures in the top of the New Pit Chalk and basal Lewes Nodular Chalk. In particular, Chalk Rock with its glauconitised hardgrounds is magnificently displayed.

Additionally, there are non-geological SSSIs at Postlebury Wood [740 430], North Brewham Meadows [743 379, 735 362], Longleat Woods [795 430], Bradley Woods [790 405], Brimsdown Hill [825 390], Long Knoll [790 376], White Sheet Hill [800 350], Cogley Wood [703 345 and 705 355], Hang Wood [860 320] and Gutch Common [895 258].

THREE

Concealed formations

An interpretation of the concealed geology of the district has been made using seismic reflection data, calibrated by boreholes at Norton Ferris, Bruton, and Fifehead Magdalen (Figure 6), supplemented by magnetic and gravity data. In addition, Palaeozoic formations crop out in the Mendip Hills, north of the district, and the structures affecting them can be extrapolated southwards. Together, the data give some indication of the deep crustal structure and show marked variations in the geometry of both the pre-Permian floor and overlying Permian and Mesozoic cover.

DEEP CRUSTAL STRUCTURE

In the north of the Wessex Basin, seismic reflection data (Chadwick et al. 1983; 1989) indicate that the crust to 30 km depth can be divided into three zones (Figure 6b). The deepest of these, Zone 3, overlies prominent, discontinuous reflectors at the base of the crust, at about 10 or 11 seconds two-way travel time. Zone 3 displays well-defined, subhorizontal, laterally discontinuous, reflectors, consistent with a metamorphic lower crust or craton of older Precambrian age.

Zone 2, the middle crustal zone, from about 2 to 15 km depth, contains several well-defined, inclined seismic reflectors. These are thought to represent folded Precambrian and Palaeozoic strata near the southern edge of the Variscan Foreland.

The Precambrian part of zone 2 is interpreted as a unit of crystalline basement overlain by volcanic and sedimentary rocks. The south-dipping, basal Cambrian reflector is well defined at about 3 km depth in the anticlinal culmination just north of the district (Figure 6c) and, although not recognised with confidence, is thought to be about 8 km deep in the south of the district. Ordovician mudstones and siltstones occur at depth below the Mendips and probably occur as far south as Maiden Bradley (Chadwick et al., 1983). Early Ordovician orthoquartzites, which are widespread in Brittany and also occur as pebbles in the Lower Old Red Sandstone in South Wales and in Triassic strata of east Devon, may extend beneath the Cornubian peninsula, across the

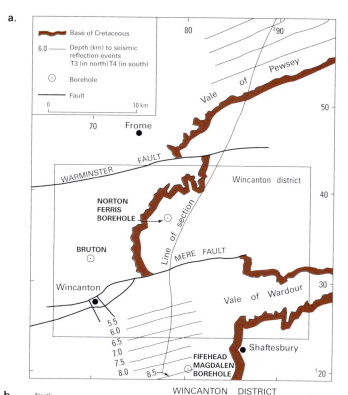

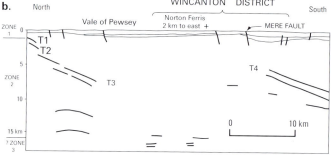

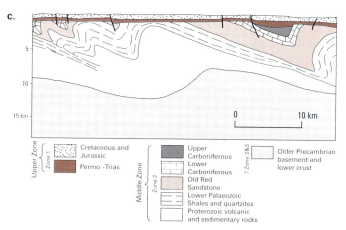

Figure 6 Location of seismic reflection data showing,
a. depth to principal inclined reflection events
b. N–S section showing principal reflectors
c. geological interpretation of reflection events (modified after Chadwick et al., 1983).

southern Bristol Channel into south Dorset (Brooks et al., 1977) and as far north as the southern edge of the Carboniferous Limestone platform just south of the district (Bristow et al., 1995, fig. 6). The upper part of Zone 2 consists of folded Silurian lavas and tuffs, Old Red Sandstone and Carboniferous Limestone (see below).

Zone 2 also contains four well-defined, south-inclined reflectors (T1 to T4, Figure 6b). Reflectors T1 and T2 are discrete events that rise to within about 1 km of the base of the Mesozoic strata, while T3 and T4 appear as a suite of reflectors, between 4 and 10 km depth. Contours on the top of T3 and T4 (Figure 6a) show that both events are planar. T3 dips at 27° toward 155° and has been mapped over an area of at least 130 km^2, T4 dips at 22° toward 165°, although it flattens beneath central Dorset (Chadwick et al., 1983).

By analogy with thin-skinned deformation typical of the Mendip Hills (Williams and Chapman, 1986), reflectors T1 to T4 are regarded as Variscan thrust surfaces or zones (Kenolty et al., 1981; Chadwick et al., 1983). Extrapolations to surface indicate that T1 to T3 subcrop beneath Mesozoic strata beneath the Vale of Pewsey, that T3 correlates with the Farmborough Fault Belt (Figure 7) and that T4 subcrops at the base of the Permo-Triassic north of the Mere Fault.

Zone 1 is highly reflective and corresponds to the Permian and Mesozoic sedimentary cover. The reflections indicate flat-lying and faulted or gently folded strata to a depth of 2 km.

SILURIAN AND UPPER PALAEOZOIC FORMATIONS

The sub-Permian basement of the Wessex Basin is divided into three east-south-east-trending tracts (Bristow et al., 1995); this district is located in the northern tract, characterised by continental-type Old Red Sandstone overlain by Carboniferous Limestone. In the central tract, south of the Fifehead Magdalen Borehole, Carboniferous Limestone is absent (Figure 7), and Devonian strata consist mainly of cleaved mudstones, transitional between continental and marine types.

In the northern tract, just north of the district, inliers of Silurian rocks, Old Red Sandstone and Carboniferous Limestone come to crop in the Mendip Hills as en-echelon periclines (Figure 7) and form large-scale, north-facing, anticlinal folds. South-dipping and folded thrusts that accompany the periclines show that the Palaeozoic strata are allochthonous (Green and Welch, 1965; Williams and Chapman, 1986). The sequences were transported north-

Figure 7 Structural features of the district and adjacent area.

A–A' Line of section (Figure 10).
B–B' Line of Section (Figure 11).

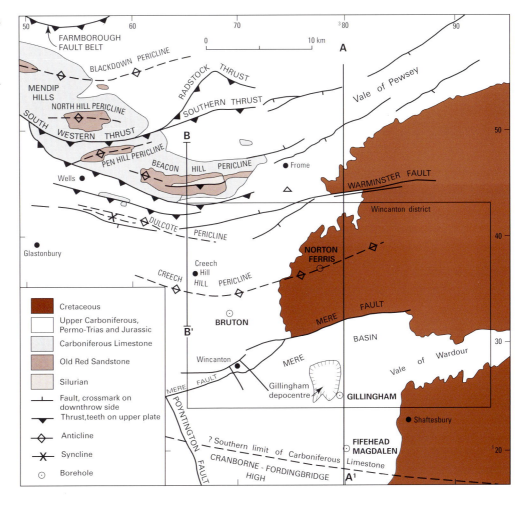

ward as a series of piggy-back thrust units during the Variscan deformation, towards the end of the Carboniferous. Restored sections suggest north–south shortening of about 40 per cent or 20 km (Williams and Chapman, 1986).

The Mendip Hills are also associated with an elongate gravity high and a linear belt of aeromagnetic anomalies (Figures 8 and 9); both trend broadly east–west as an arcuate belt, several tens of kilometres long. The anomalies are associated respectively with local outcrops of higher density Lower Carboniferous and older rocks, and of magnetic rocks of Silurian age (Ates and Keary, 1993b; Chadwick et al., 1983; Green and Welch, 1965).

Silurian rocks

The Silurian inlier in the Beacon Hill Pericline, just to the north-west of the district, was described by Hancock (1982) as an allochthonous, north-younging succession. About 350 m of strata, mainly tuffs with fossiliferous mudstones of Wenlock age, are overlain by 200 m of predominantly andesite, and capped by Old Red Sandstone. Hancock (1982) interpreted the southern margin as a Variscan thrust with moderately dipping Old Red Sandstone in the hangingwall.

The association between the aeromagnetic anomalies and Silurian andesitic lavas and tuffs (such as those which crop out at Moon's Hill [6625 4565] in the Beacon Hill Pericline) was confirmed by ground magnetic traverses (Brooks, *in* Green and Welch, 1965). A geophysical profile across the anomaly (Cornwell and Chacksfield, 1997) (Figure 10, profile A–A') shows that the lavas occur on the north side of the Pewsey Fault as a steeply folded, sheet-like body within 1 km of the surface (but also see Ates and Kearey, 1993b).

The Dulcote Pericline, which extends into the north-west of the district (Figure 7), is associated with a line of Carboniferous inliers and, like the Beacon Hill anomaly, is thought, on aeromagnetic evidence, to have Silurian volcanic rocks in its core at c. 1.5 km depth. The Carboniferous strata are deformed by a large-scale asymmetric fold with an overturned and thrusted northern limb and a more gently inclined southern limb.

Detailed ground magnetic measurements by Brooks (*in* Green and

Welch, 1965) suggested that short-wavelength anomalies, superimposed on the main anomaly, are due to source rocks at a shallow depth. Resurvey of parts of the original profiles indicate that these irregularities can be attributed to man-made sources (Cornwell and Chacksfield, 1997).

The most recent geophysical models (Cornwell and Chacksfield, 1997) indicate that magnetic Silurian lavas are present about 1.5 to 2 km below OD in the north-west of the district. The gravity data show that the Dulcote Pericline is associated with a broad, shelf-like extension of the Mendip Hills gravity anomaly. The assumed density contrast for the Mesozoic strata suggests that the base of the Permo-Triassic strata over much of this area is above OD. South of the Dulcote Pericline, magnetic anomalies disappear abruptly and suggest that either magnetic rocks are absent or they are present at a depth in excess of 5 km (Figure 11).

Devonian rocks

Only the Portishead Beds (Upper Old Red Sandstone), consisting of terrestrial sandstone with thin beds of shale, occur in the cores of the Mendip periclines (Green, 1992). Hancock (1982) described a slight angular unconformity between the near-vertical Portishead Beds and Silurian strata in the north limb of the Beacon Hill Pericline.

Devonian strata were proved in the Norton Ferris Borehole, where the composite log shows 58 m of conglomerate, sandstone, sand, siltstone and chert below the Mercia Mudstone Group. The absence of Carboniferous rocks

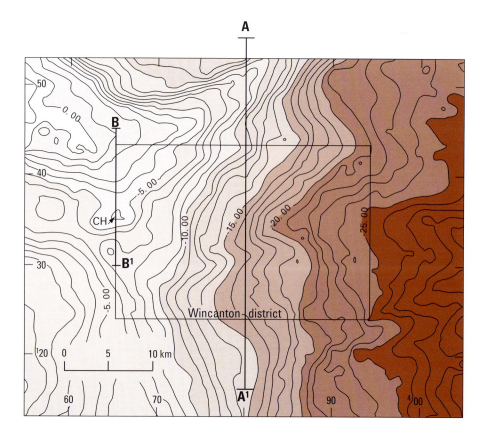

Figure 8 Bouguer gravity anomaly map of the district and adjacent area. Contours at 1 mGal intervals.
A = Creech Hill gravity high.
A–A' Line of section (Figure 10).
B–B' Line of Section (Figure 11).

Figure 9 Aeromagnetic anomaly
map of the district and adjacent
area.
Contour interval: Below -50nT = 5nT
                              Above -50nT = 25nT
A–A′ Line of section (Figure 10).
B–B′ Line of Section (Figure 11).

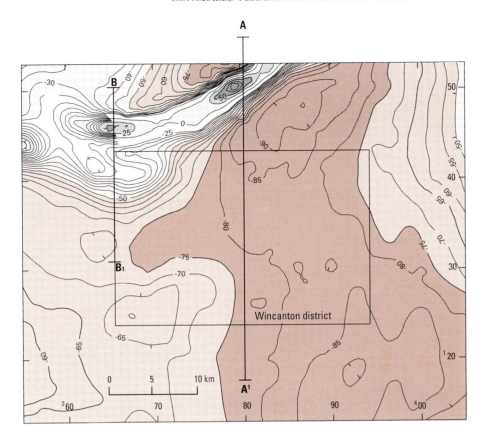

suggests that the borehole may be
sited over an antiformal thrust culmination or pericline. In the Fifehead Magdalen Borehole, 25 m of
Upper Devonian strata, here correlated with the Portishead Beds, were
proved.

Carboniferous rocks

Lower Carboniferous (Dinantian)
strata in the eastern Mendips are over
1 km thick (Kellaway and Welch,
1993). They comprise the Lower
Limestone Shale, about 150 m of dark
grey mudstones with thin limestones,
overlain by massive platform carbon-
ates of the (in ascending order) Black
Rock, Clifton Down and Hotwells
groups; dip is between 20 and 50°
south.

Late Carboniferous rocks consist of thin Namurian
strata, the Quartzitic Sandstone Formation, overlain by
Coal Measures, including sandstones of the Pennant Formation (Kellaway and Welch, 1993). These strata, over
1600 m thick, occur in synformal structures such as the
Radstock Basin, north of the Mendip periclines, and possibly beneath the Mere Basin (Figure 7), but elsewhere
are cut out by faulting.

The Bruton Borehole (Figure 7) was sited above an area
of shallow basement. It proved 92 m of early Carboniferous strata. Between 299 and 306 m depth, these rocks dip
about 20° towards 195° (Holloway and Chadwick, 1984).
The sequence is divided into an upper, interbedded limestone, siltstone and mudstone unit, about 52 m thick, and
a lower, more massive and dolomitised unit. The fauna,
which includes brachiopods, crinoids, corals and foraminifera, indicate ages in the range Arundian to Asbian
in the upper part of the sequence (Hotwells Group and
possible Clifton Down Group).

A local gravity high (Figure 8, anomaly A) some 4 km
north-west of the Bruton Borehole suggests that the basement of Carboniferous Limestone rises closer to the surface in the Creech Hill area. There, additional gravity
data (Cornwell and Chacksfield, 1997) confirmed the
position and form of the anomaly. Structure contours on
the basement (see inset map Sheet 297 Wincanton),
show that the Creech Hill anomaly is consistent with the
surface of the basement (probably Carboniferous Limestone as at Bruton) rising close to, or above, 50 m below
OD.

The profile B–B′ (Figure 11) indicates that, like the
Dulcote anomaly 5 km to the north, the Creech Hill
gravity high reflects a concealed pericline, here termed
the **Creech Hill Pericline**. The steep gravity gradients to
the south, east and north-east of it suggest that it may be
fault-bounded. In addition to a culmination of Carboniferous Limestone, higher-density Silurian or older rocks
at the concealed basement surface, could contribute to
the gravity anomaly, but the absence of a magnetic signature precludes Silurian lavas. The periclinal axis may
extend east to near the Norton Ferris Borehole, where
Devonian rocks beneath the Mesozoic succession suggest
an anticlinal fold or structural duplication.

Just south of the district, the Fifehead Magdalen Borehole proved 352 m of Carboniferous Limestone that consists of Lower Limestone Shale overlain by the Black Rock
Group and possible Clifton Down Group.

**CONCEALED PERMIAN AND MESOZOIC
FORMATIONS**

A combination of surface mapping, seismic data and boreholes allow a detailed interpretation of the structural geometry of the Permian and Mesozoic cover. Marked thickness changes, particularly of Permo-Triassic and Jurassic
strata, are a result of sedimentation across syndepositional
normal faults, such as the Vale of Pewsey and Mere faults
(Figure 6b). Both faults overlie south-dipping seismic
reflectors interpreted as thrust zones at depth, which were

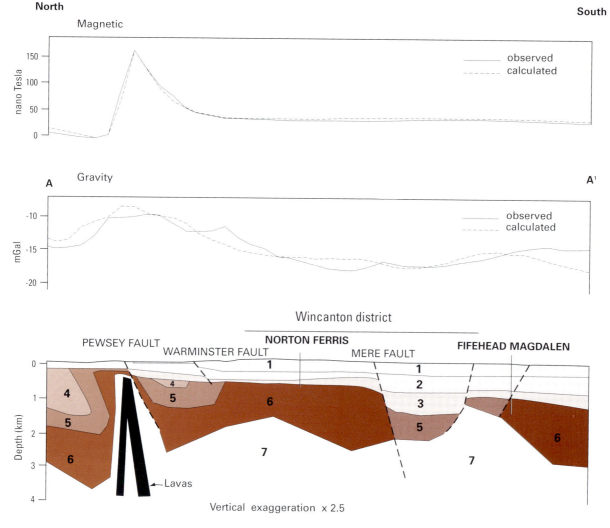

Figure 10 Observed aeromagnetic and Bouguer gravity profile A–A' along National Grid line ³80E (for location see Figure 7). The following density values (Mg/m³) and magnetic susceptibilities(SI units) were used.

1	Cretaceous 2.25; 0.0	4	Westphalian 2.60; 0.00
2	Jurassic 2.45; 0.00	5	Dinantian 2.65; 0.00
3	Permo-Triassic 2.55; 0.00	6	Devonian 2.660; 0.00

7 Silurian volcanic rocks 2.70; 0.025
8 Other basement rocks 2.70; 0.00

reactivated in part during Mesozoic extension. The most recent, Neogene, deformation occurred during a period of compressive reactivation. The surface and subsurface structure of the Mere Fault, and its displacement history are described in Chapter 8.

The Warminster Fault also shows syndepositional movement during part of the Mesozoic, and was similarly reactivated in compression during the Neogene (see Chapter 8).

In conjunction, the structure-contour maps of the pre-Permian floor (see inset map Sheet 297 Wincanton) and top Penarth Group (Figure 12) show large thickness variations in the Permo-Triassic strata. The Gillingham depocentre (Barton et al., 1998) forms a deep circular depression in the Mere Basin, 10 km across and more than 1700 m below OD in its centre (Figure 7). By contrast, the top of the Palaeozoic formations is less than 500 m below OD across much of the north-west of the district and above OD in the far north-west, indicating almost 2 km of topographical relief on the sub-Permian floor. The regional dip of subcrop surfaces is toward the east (Figure 12).

Permo-Triassic, undivided

Interpretation of seismic data suggests that the Gillingham depocentre contains 350 m of strata below the Mercia Mudstone Group (Bristow et al., 1995; Evans and Chadwick, 1994). This shows a 'blank' seismic signature that is typical of the Sherwood Sandstone Group, but there may be some Permian strata at the base.

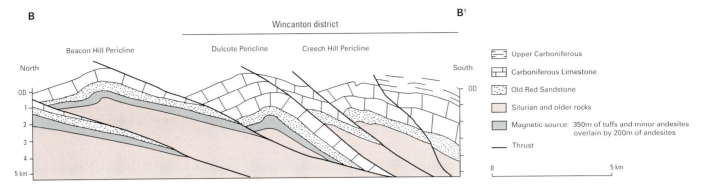

Figure 11 Interpretative cross-section B–B' along grid line ³65E (see Figure 7) showing thin-skinned deformation in the Variscan basement and constrained by ground and aeromagnetic profiles and models by Chacksfield (1995) (with data from Williams and Chapman, 1986).

Triassic

The Triassic succession is divided into four groups, in ascending sequence, the Aylesbeare Mudstone, Sherwood Sandstone, Mercia Mudstone and Penarth groups. The combined thickness of these groups as proved in boreholes ranges from 24.5 m at Bruton, 165.5 m at Norton Ferris and 182 m at Fifehead Magdalen. These thicknesses contrast with 750 m of Permo-Triassic strata in parts of the Mere Basin.

Triassic strata onlap northwards onto the Mendip High; only the upper units, the Mercia Mudstone and Penarth groups, are present north of the Mere Fault, and these die out to the north of the district. Seismic reflection data indicate similar onlap onto the Cranborne–Fordingbridge High (for location, see Figure 7) to the south of the Mere Basin. Final submergence of the Cranborne–Fordingbridge High occurred during the late Triassic, while the Mendip High survived into Jurassic times.

Aylesbeare Mudstone and Sherwood Sandstone groups

Neither the Aylesbeare Mudstone Group nor the Sherwood Sandstone Group has been proved in boreholes within the district. The nearest proving is in the Mappowder Borehole some 20 km to the south, where 184 m of Aylesbeare Mudstone is overlain by 209 m of Sherwood Sandstone. However, in the Mere Basin, seismic interpretation suggests that some 350 m of post-Variscan strata may occur beneath the base of the Mercia Mudstone Group. Whilst much of this interval has a seismic signature typical of the Sherwood Sandstone Group, regional considerations suggest that the Sherwood Sandstone is unlikely to be 350 m thick. Thus the Aylesbeare Mudstone Group could be present in the Mere Basin.

Mercia Mudstone Group

There are large changes in thickness of Mercia Mudstone Group, both in and outside the district. Just to the north, the group is absent or seen only in fissures in Carboniferous Limestone. Some 14.5 m were proved in the Bruton Borehole and 158 m in the Norton Ferris Borehole. South of the Mere Fault, seismic reflection data suggests that

the Mere Basin contains about 400 m of strata correlated with this group. Farther south, the Fifehead Magdalen Borehole, sited on the Cranborne–Fordingbridge High on the south side of the Mere Basin, proved 180 m of Mercia Mudstone Group. Geophysical logs indicate that the thin successions at Norton Ferris and Fifehead Magdalen correlate with the upper part of the Mercia Mudstone Group in the Mappowder Borehole farther south, consistent with onlap onto the Mendip and Cranborne–Fordingbridge highs respectively. The depositional environment of the Mercia Mudstone Group is interpreted as a playa or inland sabkha (Allen and Holloway, 1984).

The Mercia Mudstone Group consists predominantly of reddish brown, silty to sandy mudstone with minor interbeds of siltstone and sandstone. In the Winterborne Kingston Trough of south-central Dorset (Bristow et al., 1995, fig. 6), beds of halite up to 170 m thick occur in the middle of the sequence. The uppermost strata, the Blue Anchor Formation (formerly the Tea Green Marl and Grey Marl) consist of grey to green calcareous siltstone and mudstone. Halite is absent in the Bruton and Fifehead Magdalen boreholes, but the Norton Ferris Borehole proved four units of anhydrite within a succession of pale green and red mudstone and siltstone. The Bruton Borehole proved 6.6 m of green to grey mudtone with interbeds of muddy calcareous siltstone and limestone (Blue Anchor Formation) and 7.9 m of red silty mudstones with traces of anhydrite. The Blue Anchor Formation is 7.3 m thick in the Norton Ferris Borehole and 12.2 m thick in the Fifehead Magdalen Borehole, but is over 60 m in the Winterborne Kingston Trough to the south.

Penarth Group

The Penarth Group occurs across the district and comprises the Westbury Formation overlain by the Lilstock Formation. The upper part of the Lilstock Formation, the Langport Member (formerly the White Lias), is a massive limestone that forms a very prominent seismic reflector, and allows the top of the Penarth Group to be mapped with confidence (Figure 12). The Penarth Group was probably deposited in marginal marine and lagoonal

Figure 12 Depth-converted map showing the top of the Penarth Group. Contours at 50 m intervals.

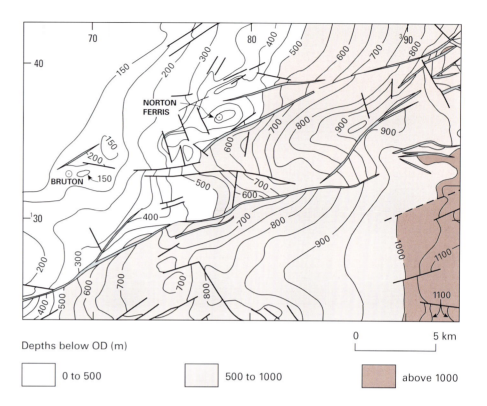

Depths below OD (m)

0 to 500   500 to 1000   above 1000

environments. It marks the close of a long period of continental, red-bed sedimentation. The group thickens southwards: it is 7.9 m thick at Norton Ferris, 10.0 m at Bruton and 18.6 m at Fifehead Magdalen just south of the district.

The Westbury Formation consists typically of dark grey to black, laminated, finely micaceous and carbonaceous mudstone. It is 1.8 to 2.1 m thick in the district.

The overlying Cotham Member of the Lilstock Formation consists, in the Bruton Borehole, of about 1.5 m of white to pale greenish grey, calcareous siltstone with traces of pyrite and shell fragments. The Langport Member comprises 6.7 m of hard, white to pale grey, sparsely bioclastic limestone with carbonaceous specks and traces of pyrite. The limestone is 4 m thick in the Norton Ferris Borehole, thickening southward to 12.8 m in the Winterborne Kingston Borehole (Knox, 1982).

SUB-CRETACEOUS FORMATIONS

A map of the strata below the Cretaceous unconformity, compiled by Chadwick and Kirby (1982) and based on seismic reflection data from the base of the Lower Greensand/Gault and the base Portland Group, can be modified as a result of this survey. The continuity of the Corallian Group below the Cretaceous is known from surface mapping, and seen as far east as both the inlier between Stourton and Penselwood, and the Norton Ferris Borehole, where Corallian Group strata were proved. The Kimmeridge Clay has a wide subcrop north of the Mere Fault; the Portland Group and Purbeck Formation occur farther north-east. The Portland Group appears to be present below Cretaceous strata in the south-east of the district, with the Purbeck Formation and Wealden Group appearing successively farther east.

FOUR

Lower Jurassic

LIAS GROUP

The Lias Group crops out in the west of the district and comprises an alternating succession of limestone and mudstone that forms part of a coarsening-upwards and shallowing-upwards sequence, with the Bridport Sands at the top. The group is dominated in the lower part by mudstone, with varying amounts of silt and thin limestone, passing up into more sandy strata and thicker limestone units. These characteristics are used to divide the Lias into lithostratigraphical units, of which several are newly recognised (Table 4). The base of the Jurassic System is taken at the lowest occurrence of *Psiloceras planorbis* in the lower part of the Blue Lias. The Blue Lias thus spans the Triassic/Jurassic Stage boundary, but for convenience, it is treated in its entirety in this chapter.

Lias strata show a general thickening southwards across the district. The group is absent north of the district on the Mendip High. There, any Lias sediments deposited across the high were largely removed by erosion in the Early Bajocian. The group is 55 m thick in a borehole [6921 4363] just south of the Cranmore Fault, 148 m thick in the Norton Ferris Borehole and 231 m in the Bruton Borehole. It is about 350 m thick on the southern margin of the district (Figure 13). It continues to thicken southwards, and 393 m were proved in the Fifehead Magdalen Borehole. The maximum thickness at outcrop is about 150 m near Bruton.

In detail, the sequence is highly variable and was affected by the syndepositional Warminster Fault. Northward thickening of Pliensbachian to Early Toarcian strata across the Warminster Fault (Chapter 8) indicates development of a half-graben at that time, controlled by uplift of the Mendip High and northward downthrow on the Warminster Fault. A change to northward thinning of the Bridport Sands across the fault indicates inversion of the half-graben controlled by reversed movement on the Warminster Fault (downthrow to the south) in the late Toarcian, either syndepositionally or postdepositionally. Movement on the Warminster Fault appears to have ceased by the Early Bajocian, as it did not affect Inferior Oolite deposition (Chapter 5). The present northerly downthrow of the fault is due to Neogene reactivation with reversed, down-to-the-north throw (Chapter 8). The Mere Fault also affected Lias deposition and a detailed account is given in Barton et al. (1998).

In the Wincanton and adjacent districts, all the Lower Jurassic ammonite zones (Table 4), except the *turneri* Zone, have been proved; some subzones are unproved due to non-exposure, others are absent due to non-sequences.

Lower Lias

The Lower Lias comprises a sequence of limestone, mudstone and sandy mudstone up to about 150 m thick. The formation, which is divisible into four members, ranges from the Hettangian to the middle of the Pliensbachian (Table 4).

BLUE LIAS MEMBER

The base is not exposed in the district; about 20 m of the upper part of the member crop out in the north-west of the district around Stoney Stratton. The member is 23.8 m at Norton Ferris and 52.3 m at Bruton [6896 3284]. It

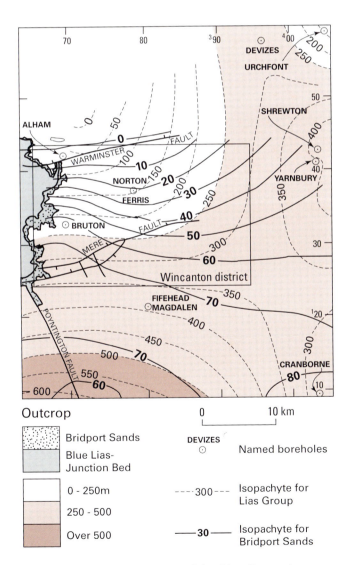

Figure 13 Isopachyte map of the Lias Group (contours at 50 m intervals) and Bridport Sands (contours at 10 m intervals).

Table 4
Classification of
the Lias Group
in the district.

Stage	Zone	Subzone	Dorset coast Cope et. al. (1980a)	Kellaway & Wilson (1944); Wilson et al., (1958)		Wincanton district	
TOARCIAN	*levesquei**	*aalensis**	Bridport Sands	Yeovil Sands	UPPER LIAS	Bridport Sands	Junction Bed
		*moorei**					
		levesquei	Down Cliff Clay			Down Cliff Clay	
		*dispansum**		'Black clay'			
	*thouarsense**	*fallaciosum**				Barrington Beds	
		*striatulum**					
	*variabilis**						
	*bifrons**	*crassum**	Junction Bed	Upper Lias Limestone			
		*fibulatum**					
		*commune**					
	*falciferum**	*falciferum**					
		*exaratum**				Marlstone Rock Bed	
	tenuicostatum	*semicelatum**					
		tenuicostatum					
		clevelandicum					
		paltum					
PLIENSBACHIAN	*spinatum**	*hawskerense*	Marlstone Rock Bed	Marlstone	MIDDLE LIAS	Marlstone Rock ? Bed	
		*apyrenum**					
	*margaritatus**	*gibbosus*	Thorncombiensis Bed	Pennard Sands		Pennard Sands	
		subnodosus	Thorncombe Sands				
		*stokesi**	Down Cliff Sands	Middle Lias Marls			
	*davoei**	*figulinum*	Green Ammonite Beds	Belemniferous and Micaceous Marls		Ditcheat Clay	
		capricornus					
		*maculatum**					
	*ibex**	*luridum**	Belemnite Stone			Spargrove Limestone	
		*valdani**	Belemnite Marls				
		*masseanum**					
	*jamesoni**	*jamesoni**					
		brevispina					
		polymorphus					
		taylori	Armatus Limestone	Lower Lias Marls			
SINEMURIAN	*raricostatum**	*aplanatum*					
		macdonnelli					
		*raricostatoides**					
		densinodulum					
	*oxynotum**	*oxynotum**	Black Ven Marls		Clays and shales	Pylle Clay	
		simpsoni					
	*obtusum**	*denotatus*					
		*stellare**					
		obtusum					
	turneri	*birchi*					
		brooki	Shales-with-beef				
	*semicostatum**	*resupinatum*					
		*scipionianum**					
		*lyra**			Clays and shale with limestones		
HETTANGIAN	*bucklandi**	*rotiforme*	Blue Lias	Blue Lias		Blue Lias	
		conybeari					
	*angulata**						
	liassicus	*laqueus*					
		portlocki			Blue Lias		
	*planorbis**	*johnstoni*					
		planorbis					

* Zone or Subzone
proved in the
Wincanton and
Glastonbury
districts

thickens southwards, and is about 136 m thick at Fifehead Magdalen just south of the district.

The Lias of the north Somerset region can be divided into three palaeogeographical domains. The *Radstock Shelf* lies to the north of the Mendip High and is not considered further in this account. The *Mendip Littoral Area*, also north of the district, is limited to the south by the Bodden Fault (Sheet 280 Wells) Sheet) and its eastwards continuation, the Cranmore Fault (Sheet 281 Frome). The *Central Somerset Basin* lies south of the Mendips (Kellaway and Wilson, 1941) and includes most of this district.

The *Mendip Littoral Area*, 1 to 3 km wide, was a near-shore, high-energy, shoal environment developed on both sides of the Mendip High. There, Lias deposits are characterised by the **Downside Stone** facies (Richardson, 1911). The Downside Stone is particularly well developed around Shepton Mallet [for example at 619 444 and 635 448]. It is up to 20 m thick and consists of massive, coarse-grained, detrital limestone, locally conglomeratic, but with no argillaceous beds. It yields abundant bivalves and gastropods, but only rare ammonites; these are of Hettangian and earliest Sinemurian (*bucklandi* Zone) age. Such sediments are commonly thicker than the equivalent beds in the Central Somerset Basin.

The **Bowlish** facies lies to the south of the Bodden and Cranmore faults and was formerly exposed just west of the district, at Bowlish [6118 4392] and near Cannard's Grave [629 416], where it is over 8 m and 7.2 m thick, respectively (Donovan, 1958b; Green and Welch, 1965). This facies of the Blue Lias consists of coarsely crystalline to fine-grained and argillaceous, bedded limestones, in beds usually less than 0.3 m thick, interbedded with thin mudstones. It is also of Hettangian and earliest Sinemurian age. Although the Bowlish facies does not crop out in the district, it probably occurs at depth on its northern margin.

The southwards transition, through the Bowlish facies into the 'normal' facies in the Central Somerset Basin is rapid (Green, 1992), perhaps occurring abruptly across the Warminster Fault. In the basin, up to 650 m of shales, thin limestones and sands represent normal deposition in a relatively deep-water, offshore, marine-shelf environment (Cope et al., 1980a; Duff et al., 1985). The basinal facies crops out in the north-west of the district. It consists of alternating muddy limestones, in beds generally 0.1 to 0.3 m thick, and thin, interbedded silty mudstones. Limestones are dominant, forming up to 80 per cent of the succession in the north-west; they commonly form long dip slopes, and are found in the beds of ditches and some of the smaller streams, for example north of Stoney Stratton [6566 3966 to 6564 3951]. A representative section in a pit [6591 3857] south of Stoney Stratton shows 3 m of interbedded biomicritic limestones, 0.1 to 0.5 m thick, with sparry patches and thinner marly to sandy clays. Rhynchonellid brachiopods are common in the limestones, together with bivalves and lignite fragments. In the more southern basinward exposures, mudstones dominate, as in the railway cutting at Pylle [6065 3870] just west of the district.

The Blue Lias ranges from latest Triassic to Early Sinemurian in age. The oldest strata to crop out in the district probably belong to the *bucklandi* Zone (Bristow and West-

head, 1993), although at Cannard's Grave [629 416], just to the west, the *angulata* and *planorbis* zones were proved (Donovan, 1958). The youngest strata probably belong to the *semicostatum* Zone, *lyra* Subzone, as proved by material in the BGS collections from just north-west of the district [6156 4048] (Ivimey-Cook, 1993a, unpublished report*).

PYLLE CLAY (MEMBER)

The type area of the Pylle Clay is the hamlet of Pylle [608 383] just west of the district (Bristow and Westhead, 1993). The member is equivalent to the Lower Lias Shales of Duff et al. (1985), the lower part of the Lower Lias Marls of Kellaway and Wilson (1941), and the Black Ven Marls and lower part of the Belemnite Marls of the Dorset coast.

The Pylle Clay crops out in the north-west of the district, around Stoney Stratton. It is 33.8 m thick in the Bruton Borehole and 40.2 m thick in the Norton Ferris Borehole, but is absent in the Alham Borehole north of the Warminster Fault. At outcrop, it consists of 15 to 20 m of dark grey mudstone. The characteristic colour is evident in both weathered and unweathered sections; a few thin muddy limestones occur. The member was exposed in the small outlier at Shepton Mallet Cemetery [619 440] north-west of the district (Green and Welch, 1965, p.104). Typical exposures [6661 3903, 6661 3895 to 6666 3839] in Fosse Combe Gully, east of Stoney Stratton, a few metres below the top of the member, reveal 1.2 m of dark grey to black, fissile mudstone grading up into paler grey, fissile, marly clays with belemnites, and sporadic, very muddy limestone, less than 0.1 m thick.

The member is richly fossiliferous. To the south (Hollingworth et al., 1990) and west (Donovan et al., 1989), the Pylle Clay includes beds that range from the *obtusum* Zone, *stellare* Subzone, to the *raricostatum* Zone, *raricosta-toides* Subzone (Table 4). At Cannard's Grave [629 416], just west of the district, the *oxynotum* Zone and Subzone were proved (Donovan, 1958), and at Prestleigh, also just west of the district, the *raricostatum* Zone was proved in brash [631 407] from a small faulted outcrop.

SPARGROVE LIMESTONE (MEMBER)

The Spargrove Limestone takes its name from the hamlet of Spargrove [672 380] where the limestone forms well-featured ground (Bristow and Westhead, 1993). The member crops out in the north-west of the district around Stoney Stratton [664 392]. In the Bruton Borehole [6896 3284], the member is 8 m thick, and at Norton Ferris, it is 5.8 m thick.

At outcrop, it comprises up to 2.5 m of interbedded muddy limestone, in beds up to 0.1 m thick, and thin silty mudstone. In the Alham Borehole [6793 4118], 7.6 m of massive limestones overlie the Blue Lias (the Pylle Clay is missing, possibly represented by an erosion surface at about 98 m depth). These limestones are probably a thick, lateral equivalent of the Spargrove Limestone, and the same as the limestones noted by Green and Welch (1965, p.103) just north-west of the district [617 449 and 605 469], on the flank of the Mendips High. The Jamesoni and Valdani limestones of the Radstock area, north of the Mendips, are an approximate equivalent of rather different lithology (Green, 1992; Tutcher and Trueman,

1925). On Pennard Hill, west of the district, the member is absent.

A rich ammonite fauna (detailed in Bristow and West-head, 1993; Westhead 1994) indicates all subzones between the *jamesoni* Zone, *jamesoni* Subzone, and *ibex* Zone, *luridum* Subzone, with most of the member falling in the latter Zone (Table 4). The *jamesoni* Zone was proved by *Uptonia?* in the lower part of the Spargrove Limestone east of Stoney Stratton [6624 3917], and by *U. jamesoni* in Fosse Combe Gully [6662 3905]. The *ibex* Zone, *masseanum* Subzone, is suggested by *Tropidoceras* sp. in a stream bed [666 386] north of Spargrove, and also north-west of Fosse Combe Gully [663 393]. There, limestone debris includes common *Acanthopleuroceras arietiforme?*, *A. maugenesti*, *A. valdani*, *Tragophylloceras loscombi* and *T. ibex*, indicative of the *ibex* Zone (mainly *valdani* Subzone, but possibly also the preceding *masseanum* and succeeding *luridum* subzones). The *valdani* Subzone is indicated by *T. ibex* and the index species from just beneath Ditcheat Clay in an exposure [6657 3926] in a swallet at the head of Fosse Combe Gully and by *T. ibex*, *A.* aff. *valdani* and *Liparoceras* sp. in Chesterblade Bottom [6588 4025]. A bed near the top of the member in the Alham Borehole yielded *Beaniceras* and *Tragophylloceras*, which suggest the *ibex* Zone, *luridum* Subzone.

DITCHEAT CLAY (MEMBER)

The type area of this unit of silty clay is the village of Ditcheat [625 362] just west of the district (Bristow and Westhead, 1993). The member equates with the upper part of the Lower Lias Marls of Kellaway and Wilson (1941), the Belemnitiferous and Micaceous Marls of Wilson et al. (1958) and the Green Ammonite Beds of the Dorset coast (Table 4).

The member crops out between Chesterblade and west of Bruton. It is about 60 m thick in the type area thinning to about 25 m south-west of Chesterblade and 24 m in the Bruton Borehole; the Norton Ferris Borehole proved 21.6 m. The member is anomalously thick, 46.95 m, in the Alham Borehole, north of the Warminster Fault, but thins rapidly northwards or is cut out beneath the unconformable Inferior Oolite. Just north of the district, boreholes [6838 4371 and 6921 4363] close to the Cranmore Fault proved 20 and 30 m, respectively.

The Ditcheat Clay consists of pale to medium grey, silty mudstone and silt, weathering to a characteristic mottled orange and grey; a few siltstone and ironstone nodules, and silty limestones, up to 0.15 m thick, occur. A thicker sequence in the Alham Borehole consists of bluish grey, silty, calcareous and micaceous mudstone, with calcareous and ferruginous concretions.

Fossils are scarce, but include *Androgynoceras* cf. *maculatum* from the old brickpit [6385 3700] north-east of Ditcheat (Kellaway and Wilson, 1941), and *Androgynoceras* sp. on the south side of Pennard Hill (Donovan et al., 1989). These ammonites indicate the *davoei* Zone, *maculatum* or *capricornus* subzones. Just east of Stoney Stratton, a section [6657 3926] with *Beaniceras crassum* and *Lytoceras fimbriatum* show the presumed basal beds of the Ditcheat Clay to belong to the *ibex* Zone, middle *luridum* Subzone. As the top of the Spargrove Limestone also falls in the

luridum Subzone, the junction between these two units is dated fairly precisely.

Middle Lias

In the present area, the Middle Lias is represented by the Pennard Sands and the lower part of the Junction Bed.

PENNARD SANDS (MEMBER)

The Pennard Sands of this account equate with the combined Pennard Sands and Middle Lias Marls of Kellaway and Wilson (1941) (Table 4). The member takes its name from Pennard Hill just west of the district. Within the district, the member crops out between Chesterblade [660 408] and Bruton, at Yarlington [650 290] and Lower Woolston [655 272]. North of the district, around Cranmore, it is absent.

The Pennard Sands vary considerably in thickness. Near Chesterblade, north of the Warminster Fault, they are only 5 to 13 m thick. In the Alham Borehole, the member is represented by 12.65 m of grey, silty to sandy, calcareous, micaceous mudstones. South of the fault, around Alham [680 405], the member varies over a short distance from 5 to 25 m thick. Farther south, towards Bruton, they vary from 20 to 40 m thick, and are 31.4 and 21 m thick, respectively, in the Bruton and Norton Ferris boreholes.

The member consists of silty to fine-grained sandy clay, clayey and silty, fine-grained sand and thin beds of fine-grained, commonly micaceous sand. Weak springs commonly mark its base. On Pennard Hill, west of the district, there is a persistent bed of fine-grained sand, up to 5 m thick, about 15 to 25 m above the base of the member. One of the best sections is a river cliff [6869 3873] south of Batcombe which reveals 6 m of dark greyish blue, fissile, sandy to silty clay.

Fossils are scarce. Donovan et al. (1989) found *Amaltheus stokesi* in the basal sand on the south-west side of Pennard Hill [5754 3729], west of the district. This indicates the *margaritatus* Zone, *stokesi* Subzone.

Middle/Upper Lias

JUNCTION BED (MEMBER)

The Junction Bed is a thin, but widespread, condensed deposit, 1 to 6 m thick, that spans the Pliensbachian/Toarcian (Middle/Upper Lias) Stage boundary and encompasses six ammonite biozones (Table 4).

It crops out in the north-west of the district. North of the Warminster Fault, the Junction Bed occurs only in thin lenses, up to 1.5 m thick, beneath unconformable Bridport Sands or Inferior Oolite; in the Alham Borehole, it is 1.1 m thick The member is well represented south of the Warminster Fault (6 m thick at Higher Alham [6816 4104] (see below) and 6.3 m at Batcombe (De La Beche, 1846)). In places, such as around Creech Hill, it is cut out beneath the Bridport Sands. The Junction Bed is 5.2 m thick in the Bruton Borehole and 3.96 m thick in the Norton Ferris Borehole.

The Junction Bed is divisible into a lower **Marlstone Rock Bed** and an upper **Barrington Beds** (Bristow and

Westhead, 1993; Donovan et al., 1989), although in the present district, it has been mapped as a single unit.

The **Marlstone Rock Bed** comprises ooidal, belemnitiferous, grey, commonly ferruginous-weathering calcarenite, with mudstone clasts and pebbles. The maximum thickness is about 1.5 m, although 3 m were recorded at Maes Down [around 645 406], just west of the district (Richardson, 1906). Fauna from Pennard Hill, west of the district, indicate that the Marlstone Rock Bed ranges from the *margaritatus* Zone up into the *tenuicostatum* Zone, *semicelatum* Subzone (Table 4).

The **Barrington Beds** (Donovan et al., 1989), the Upper Lias Limestone of Kellaway and Wilson (1941), comprise up to 3 m of sparsely ooidal micrite and calcarenite with common ammonites and rare belemnites. They span the *falciferum* to *thouarsense* zones.

A representative section [6638 4047] is exposed on the west side of Small Down; this is almost certainly Richardson's (1909b, p.541) 'Small-Down Quarry, near Evercreech'. The section exposes 1 m of iron-stained biomicrite with rusty peloids and washes of current-aligned belemnite guards, overlying 2 m of massive, uneven, slightly sandy biomicrite with sparry patches and rare peloids. The lower 2 m is almost certainly the Marlstone of Richardson (1909b). His bed 5,'fitting into the irregularities of the Marlstone', probably belongs to the *tenuicostatum* Zone (not '?Falciferi' as he supposed). Fossils in the rest of the section indicate the *exaratum* (beds 4, 3c), *fibulatum* or *crassum* subzones (bed 3a), and *crassum* Subzone or *variabilis* Zone (bed 1).

In the Alham Borehole, the Junction Bed is 1.1 m thick and comprises 1 m of argillaceous, bioturbated limestone or silty, calcareous, micaceous mudstone, overlying 0.1 m of ooidal micrite with clasts 20 to 30 mm across of non-ooidal carbonate and an erosional base.

A section at Higher Alham [6816 4104] exposes, in upward succession, 2.5 m of coarse biosparite with rusty patches and some peloids in 0.1 to 0.4 m-thick beds: 1 m to 2 m of very uneven, rusty, finer-grained sparite with common shell fragments and large belemnite guards: 1 m of uneven-bedded micrite with common belemnites, bivalves and ammonites: and 1 m of irregular, shelly micrite with rusty peloids, belemnites and ammonites, including *Dactylioceras* sp. and *Hildoceras bifrons* indicative of the *bifrons* Zone, *fibulatum* or *crassum* subzones.

The local range of the Junction Bed is from the *spinatum* Zone, *apyrenum* Subzone, to the *thouarsense* Zone, *fallaciosum* Subzone (Table 4), but there is no direct evidence of the *tenuicostatum* Zone. This range was proved by the rich fauna collected from brash of ironstained, peloidal, dark grey, biomicritic limestone at Westcombe [around 6718 3923] (Table 5).

Elsewhere, the *falciferum* Zone and Subzone and the *bifrons* Zone, *commune* Subzone, are

indicated by *Harpoceras* sp. and *Hildoceras lusitanicum* in micrite with rusty peloids in patchy exposures [6787 4087] west of Higher Alham. The *bifrons* Zone, *fibulatum* or *crassum* Subzone, was proved by ammonites in brash [678 391] of sparry to micritic, porous, ferruginous limestone around a pit at Westcombe (Bristow and Westhead, 1993). This is probably the pit where Richardson (1909b, p.542) recorded 'Marlstone', overlain by 0.61 m of 'Toarcian ironshot beds'. Brash [6674 4010] south of Small Down Farm yielded: *Hildoceras laticosta*, indicative of the *falciferum* or *commune* subzones, in grey, ironshot limestone: *Hildoceras bifrons*, indicative of the *fibulatum* or *crassum* subzones, in pale grey, fine-grained, strongly ironshot limestone: and *Haugia variabilis*, zonal index of the *variabilis* Zone, in either fine-grained limestone or impure, ochreous limestone.

Upper Lias

The Upper Lias comprises two members: thin, argillaceous Down Cliff Clay at the base, capped by thicker, arenaceous Bridport Sands.

DOWN CLIFF CLAY (MEMBER)

The Down Cliff Clay takes its name from the cliffs between Ridge (Down) Cliff and Thorncombe Beacon on the Dorset coast (Buckman, 1922). In the Wincanton district, it crops out south of the Warminster Fault as far south as Milton Clevedon. It consists of sandy to silty, micaceous, sparsely bioturbated mudstone similar to the Pennard Sands. In places, it is cut out beneath Bridport Sands. North of the Warminster Fault, the member was proved only in the Alham Borehole where it is an anomalous 11.63 m thick. The base is seen in a river-cliff [6901 3876] south of Batcombe, where 1 m of sandy clay overlies 2.15 m of Junction Bed. Sections in the middle of the member in a stream [6936 3975] near Batcombe, show dark grey, clayey, sandy silt with a thin (0.15 m) bed of sandstone. The upper boundary was seen in a nearby stream-bank [6929 3991] where 1 m of sand (Bridport Sands), overlies 1 m of clayey sand.

The Down Cliff Clay probably belongs to the *dispansum* Subzone (Table 4).

Table 5 Junction Bed fossils from field brash at Westcombe [6718 3923].

AMMONITE	ZONE	SUBZONE
Pseudogrammoceras	*thouarsense*	*fallaciosum*
Grammoceras thouarsense	*thouarsense*	*striatulum*
Esericeras sp.	*thouarsense*	*striatulum*
Haugia variabilis	*variabilis*	
Phymatoceras binodatum	*variabilis*	
Hildoceras semipolitum	*bifrons*/early *variabilis*	*crassum*
Dactylioceras cf. *athleticum*	*bifrons*	*commune*
Hildoceras laticosta	*falciferum/bifrons*	*falciferum/commune*
Harpoceras falciferum	*falciferum/bifrons*	*falciferum/commune*
Nodicoeloceras crassoides	*falciferum*	*exaratum/falciferum*
Cleviceras exaratum	*falciferum*	*exaratum*
Dactylioceras sp. indet.	*tenuicostatum/bifrons*	
Pleuroceras solare	*spinatum*	*apyrenum*
Leptaleoceras sp.	*spinatum*	

BRIDPORT SANDS (MEMBER)

Fine-grained silty sand and calcareous sandstones, principally of mid to late Toarcian age, with some of earliest Aalenian age, crop out in an arc through western Dorset and south Somerset. They form part of a coarsening-upward and shallowing-upward sequence from the Down Cliff Clay into the Inferior Oolite (Bryant et al., 1988; Holloway, 1985). Within this tract, the sands and sandstones have been variously named (Arkell, 1933). Following Arkell, the term Bridport Sands, rather than the Yeovil Sands of Hudleston (in Buckman, 1879), is used in this account for the sands between the Down Cliff Clay and Inferior Oolite.

The member crops out in the west of the district and commonly forms the steep slopes of scarps capped by Inferior Oolite. The most extensive outcrops are south of the Warminster Fault towards Bruton, and south-westwards towards Hadspen [657 326].

The thickness of the Bridport Sands is shown in Figure 13. At Chesterblade, north of the Warminster Fault, the sands are present only in lenses, up to 3 m thick, beneath unconformable Inferior Oolite; they are absent in the Alham Borehole and farther north near Cranmore. South of the fault, the Bridport Sands show a general southwards thickening; they are 15 m thick adjacent to the fault, 19.8 m in the Norton Ferris Borehole, 62 m in the Bruton Borehole and 76.2 m in the Fifehead Magdalen Borehole.

The base of the Bridport Sands, which is probably transitional over a metre or so, is taken at the incoming of silt and sand above mudstones of the Down Cliff Clay. In places, such as west of Creech Hill [around 665 360], the base of the Bridport Sands is unconformable, cutting out the Down Cliff Clay and Junction Bed.

Characteristically, yellow and orange-weathering, friable, silty, fine- and very fine-grained sandstone is in rhythmic alternation with more calcareous sandstone doggers; sandy bioclastic limestone occurs locally. One of the more extensive sections is a cutting [6888 3926 to 6895 3920] at Batcombe showing about 20 m of fine-grained, silty, orange-weathering sands, with sporadic, 0.20 m-thick, calcareous sandstone beds. Another good exposure [6704 3555], south of Creech Hill, shows 8 m of fine-grained, yellow, silty sand with 0.25 m-thick calcareous sandstone beds. The average grain-size is just above 0.6 mm, the borderline between sand and silt.

Three principal rock types occur in the Bridport Sands (Davies, 1969; Davies et al., 1971; Knox, 1982; Barton et al., 1993). The dominant lithology, in units up to 2 m thick, is olive-grey to greyish green, poorly cemented, micaceous, commonly bioturbated and generally poorly sorted, silts and very fine-grained to fine-grained sandstones. The sands are composed of angular to very angular quartz (65 to 75 per cent), mica (up to 10 per cent), feldspar (1 to 25 per cent) and a variable amount of bioclastic debris, set in a variably clayey, micritic and fine-grained, sparry calcite matrix. Cemented sandstones, up to 0.5 m thick, occur at some levels within the succession. They consist of abundant shell fragments and scattered fine sand grains in a matrix of fine-grained carbonate cement. Both matrix and bioclastic fragments are domi-

nantly of ferroan calcite. Small-scale cross- and festoon-bedding have been recorded; their absence in most sections may be due to bioturbation. At the surface, the cement is concentrated in horizontal layers up to about 0.5 m thick forming 'doggers'; some doggers mark the base of erosional structures. Davies (1967) concluded that their development was due to variations in the supply of detrital material, the doggers representing periods of reduced supply of sand and silt. Bioclastic limestones generally form a minor part of the sequence and consist primarily of broken, abraded and bored bivalve fragments, with a very small component of chamositic or phosphatised ooids and fine-grained sand. South-west of the district, an exceptional development of these limestones is the 30 m-thick Ham Hill Stone north of Crewkerne. In the Winterborne Kingston Borehole [SY 8470 9796], the clay-mineral assemblages are dominated by iron-rich chlorite, with minor amounts of mica and kaolinite. The proportions are reversed in approximately the lower third of the formation (Knox, 1982).

A wide variety of depositional models have been proposed for the Bridport Sands; these include sandbanks or shoals in shallow water (Kellaway and Welch, 1948; Bryant et al., 1988), a migrating barrier bar (Davies, 1969; Davies et al., 1971), a storm-dominated deposit in the lower shoreface environment (Colter and Havard, 1981), low-energy deposits in a shallow sea with the coarsening-upward cycles representing shoaling (Knox, 1982a), and as mixed fair-weather sediments and storm deposits (Kantorowicz et al., 1987). Large-scale cross-bedding and lack of bioturbation in the bioclastic limestones suggested to Davies (1969) rapid emplacement by high-energy currents in a tidal channel, and to Jenkyns and Senior (1991), a fault-bounded high where carbonate could accumulate.

The heavy minerals may indicate a south-westerly (Boswell, 1924), or a north-easterly (Morton, 1982) provenance. Cross-bedding measurements between the Dorset coast and the Mendips suggest transport dominantly from the north-east, with only a minor south-westerly component (Davies, 1969). Large-scale cross-bedding in the Ham Hill Stone suggests transport mainly from the west or south-west (Davies, 1969). Magnetic fabric studies are inconclusive and indicate deposition by north-easterly or south-westerly directed currents (Hounslow, 1987). Together, the data suggest both north-easterly and south-westerly source areas.

In the Winterborne Kingston Borehole [SY 8470 9796], the basal Bridport Sands, belonging to the *levesquei* Zone and Subzone, rest on Down Cliff Clay belonging to the *levesquei* and *dispansum* subzones (Ivimey-Cook, 1982). Very fine-grained, doggery sand, a few metres below the top of the member near Wyke Champflower [6682 3460], yielded *Phlyseogrammoceras?* and *Pseudogrammoceras?* indicative of a position close to the *thouarsense/levesquei* zonal boundary. There, the higher part of the *levesquei* Zone may have been removed by pre-Inferior Oolite erosion. The *dispansum* Subzone is indicated by *Phlyseogrammoceras dispansum* in a nodule from a cutting [656 266] on the Wincanton Bypass and from ammonites at the top of the Bridport Sands in the railway cutting [682 345] at Lusty. Elsewhere, however, a slightly younger, *moorei*, Subzone is suggested by *Dumortieria* aff. *pseudoradiosa* collected from a bank [6735 3318] south-west of Bruton and by *Catulloceras psamminum, Dumortieria subsolaris, D.* aff. *lata* and *Hudlestonia serrodens* from a section south of Pitcombe [677 321] (Freshney, 1994; Richardson, 1916). In the Wincanton Borehole [7116 2816], the youngest, *aalensis,* Subzone may be present (Ivimey-Cook, 1992a, unpublished report*), as in Stowell [6849 2180] (Pringle, 1910) and Purse Caundle [7012 1826] (Barton et al., 1993) boreholes south of the district. Together, these data show that where the Bridport Sands succession is complete, as in the south of the district, the top lies close to the Toarcian/Aalenian stage boundary. Elsewhere, some or all of the member has been removed by erosion prior to the deposition of the Inferior Oolite.

FIVE

Middle Jurassic (part)

Middle Jurassic strata crop out in the west of the district and comprise an alternating sequence of limestone and mudstone. The formations and members of the Middle Jurassic are shown in Tables 6 and 7.

INFERIOR OOLITE (FORMATION)

The lithostratigraphical classification and nomenclature of the Inferior Oolite of the district is based on Parsons (in Cope et al., 1980b) and Barton et al. (1993), largely following Richardson (1907; 1916). South of the Mere Fault, the stratal units correlate with those established by Barton et al. (1993) in the Purse Caundle Borehole [7012 1826] south of the district, which provides a standard for the Shaftesbury district (Table 6). The units defined between Hadspen [around 66 31] and Doulting [646 424] correlate only on a broad scale with the Purse Caundle sequence.

Over much of the district, only Upper Inferior Oolite of Late Bajocian and Early Bathonian age is preserved. Aalenian and Early Bajocian rocks occur in the Cole 'Syncline' (Richardson, 1916) (Figures 14 and 15). Although the terms Lower, Middle and Upper Inferior Oolite (Arkell, 1933) (Table 6) are referred to in this account, it is generally difficult to correlate their boundaries with those of the units described below, none of which has been mapped. Most of the Inferior Oolite of the district is Upper Inferior Oolite, and has been divided into **Garantiana Beds** and **Hadspen Stone** below and **Doulting Stone** and **Crackment Limestones** above. The **Rubbly Beds**, the lateral equivalent of the Doulting Stone may be present in the extreme south of the district. Locally, a higher unit, the **Anabacia Limestone,** occurs.

The Inferior Oolite usually caps a major escarpment and forms extensive easterly or south-easterly facing dip slopes in the west of the district. At depth, the formation dips generally eastwards to reach a maximum depth of

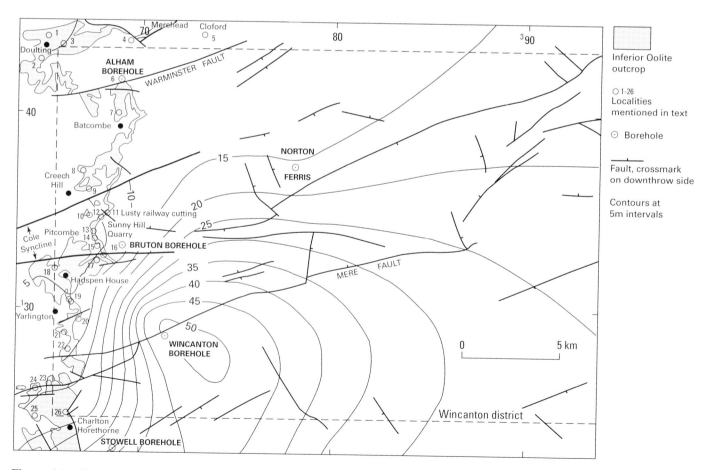

Figure 14   Outcrop and isopachyte map of the Inferior Oolite (contours at 5 m intervals) showing localities referred to in the text.

Table 6 Classification of the Inferior Oolite; and comparison with the Shaftesbury district.
The zonation follows Parsons *in* Cope et al. (1980b).

		CHRONO/BIOSTRATIGRAPHY		LITHOSTRATIGRAPHY		
STAGE	SUB-STAGE	ZONE	SUBZONE	SHAFTESBURY DISTRICT (Purse Caundle Borehole, Barton et al., 1993)	INFERIOR OOLITE FORMATION	WINCANTON DISTRICT
BATHONIAN	LOWER	*zigzag*	*yeovilensis*	FULLER'S EARTH		
			macrescens	Crackment Limestones	UPPER	
			convergens	?		Anabacia Limestone ? Crackment Limestones
BAJOCIAN	UPPER	*parkinsoni**	*bomfordi*	Rubbly Beds		Doulting Stone | Rubbly Beds?
			*truellei**			
		*garantiana**	*acris**			Hadspen Stone | Garantiana Beds
			tetragona	Sherborne Building Stone		
			subgaranti			
			dichotoma	?		
		*subfurcatum**	*baculata*			
			polygyralis		MIDDLE	
			*banksi**	Miller's Hill Beds		Doulting Conglomerate
	LOWER	*humphriesianum**	*blagdeni**			12^1–Bed 4
			humphriesianum			
			romani			
		*sauzei**				
		*laeviuscula**	*laeviuscula**			Pecten Bed
			*ovalis**			Limestone and marl
		*discites**		Corton Denham Beds		Ammonite Bed
AALENIAN		*concavum**	'Graphoceras formosum horizon'		LOWER	
			concavum			
		*murchisonae**	'Brasilia gigantea horizon'	Ringens Bed		
			bradfordensis			
			*murchisonae**			Conglomerate
			haugi			
		opalinum	*scissum*	?		
			opalinum			

* Zone/subzone proved in the Wincanton district
1 Locality 12 — see Figures 14 and 15

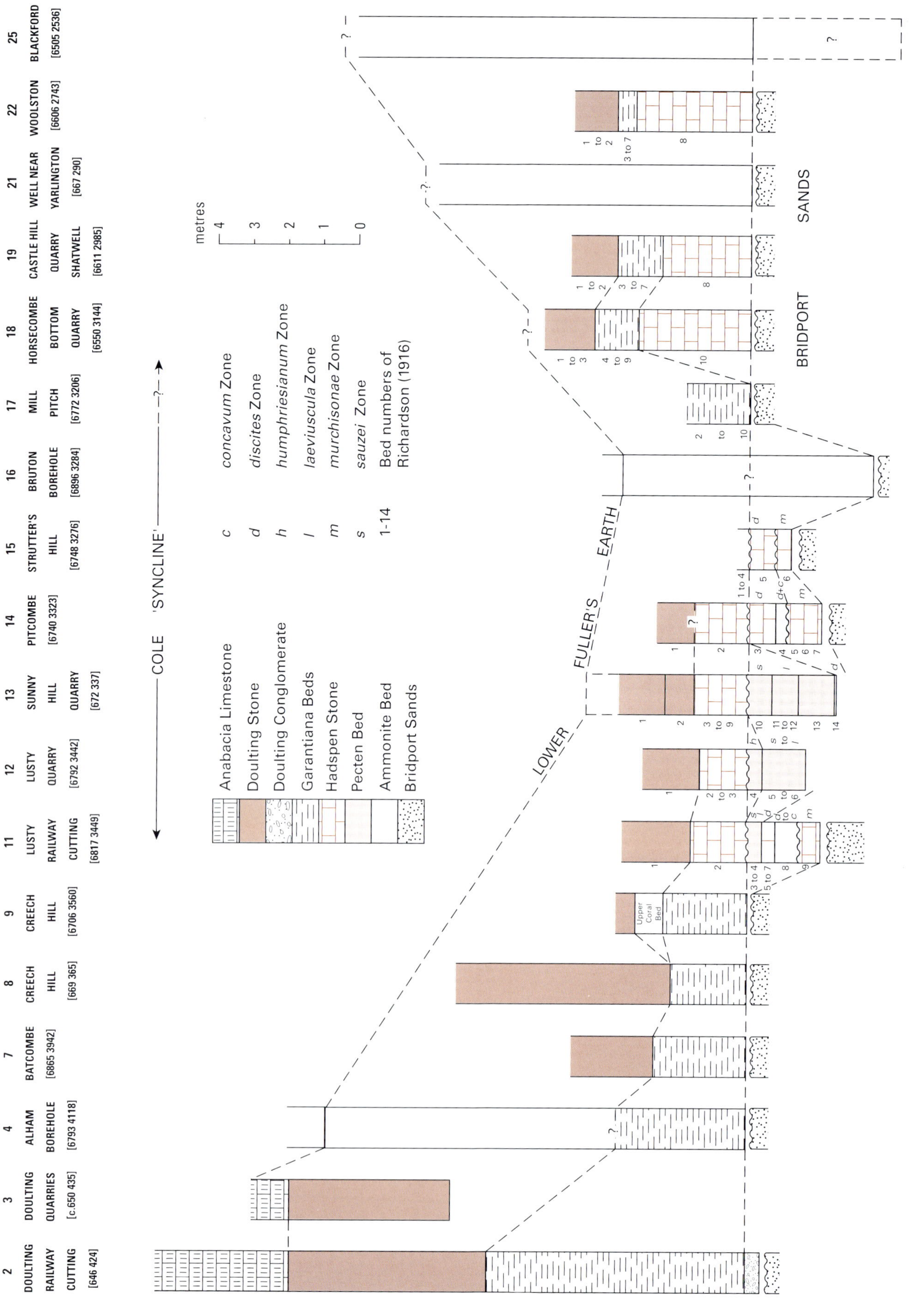

Figure 15 Sections in the Inferior Oolite across the district (see Figure 14 for location).

about 700 m below OD in the east of the district (Figure 14). The formation comprises shallow-water limestones and thin mudstones of Aalenian, Bajocian and Early Bathonian age (Table 6).

Although apparently concordant with the Bridport Sands, the base of the Inferior Oolite is commonly pebbly and marks a substantial non-sequence. Richardson (1916) recorded a generally progressive overstep by the Upper Inferior Oolite, northward from the Shaftesbury district, until it rests on Carboniferous Limestone in the Mendip Hills north of the district. This northward overstep is interrupted by the Cole 'Syncline' (Richardson, 1916), south-west of Bruton (Figure 14). The Cole 'Syncline' is probably fault-controlled and is more likely to be a shallow, east-trending, graben or half-graben. Lower and Middle Inferior Oolite are recognised around Doulting just to the north-west of the district, in the Cole 'Syncline', and in the Wincanton Borehole. In the Cole 'Syncline', there are up to 2.3 m of limestones of Late Aalenian to Early Bajocian age (**Corton Denham Beds** and **Miller's Hill Beds**) separated from both the Upper Inferior Oolite and the Bridport Sands by erosion surfaces (Table 6). In the Wincanton Borehole, the lowest 11 m (dip corrected) of Inferior Oolite probably belong to the *concavum* and *murchisonae* zones (Ivimey-Cook, 1992a, unpublished report*). The youngest Inferior Oolite (*zigzag* Zone) probably occurs only in the thicker sequences in the far north (Anabacia Limestone) and south (Crackment Limestones) of the district and adjacent areas (Bristow et al., 1995; Cope et al., 1980b); Richardson (1916, table) suggested that they might occur just west of the district. The *zigzag* Zone has been proved in the Doulting area (Savage, 1977), west of Charlton Horethorne (Taylor, 1990), and Purse Caundle Borehole (Barton et al., 1993) just south of the district.

The Inferior Oolite varies markedly and irregularly in thickness across the district. It is 16.4 m thick at Doulting (Savage, 1977) just west of the district; 11.6 m in the Alham Borehole (Figure 14, locality 6) [6793 4118]; about 12 m at Batcombe [693 385]; 5 to 6 m on Creech Hill; 15.2 m in the Norton Ferris Borehole [7820 3700]; 6.8 m in the Bruton Borehole [6896 3284]; 6 m east (Figure 14, locality 20) [667 290] and 4.5 m north-east (Figure 14, locality 19) [6611 2985] of Yarlington (Richardson, 1916); 6 m at Blackford [660 260]; 16 m just west of the district [6505 2536 and 6509 2536] (Figure 14, locality 24), and 8.6 m in a well (Figure 14, locality 21) [6595 2820] south of Yarlington. In the Wincanton Borehole [7116 2816], just south of the Mere Fault, the formation is 52 m thick (after correction for dip). At outcrop south of the Mere Fault, the formation is about 15 m thick and continues to thicken southwards; it is 36.3 m in the Stowell Borehole (Figure 14) and 44.2 m thick in the Purse Caundle Borehole in the north of the Shaftesbury district.

Palaeogeography

In south Somerset and north Dorset, the Inferior Oolite was deposited on a basinal, offshore, limestone ramp, periodically invaded by coarse, bioclastic or peloidal material, and affected by contemporaneous fault move-

ments and variation in subsidence rate (Barton et al., 1993). The exact palaeogeography of the basin, however, is unclear. In the north, the influence of the Mendip axis caused erosion or non-deposition, throughout the Aalenian and most of the Bajocian, but also continuing into the Early Bathonian. The axis may periodically have emerged; its final submergence began early in the Late Bajocian (*subfurcatum* Zone), and was followed by a more major transgression (*garantiana* Zone). The axis then became relatively inactive, although southerly palaeocurrents noted in the Doulting quarries [650 435] suggest that it may still have acted as a passive high area during *parkinsoni* Zone times.

Faulting, particularly during the Aalenian and Early Bajocian, influenced both lithology and preservation (Barton et al., 1993). At outcrop, the best developments of these strata lie in the Cole 'Syncline'; thick equivalents of the Miller's Hill and Corton Denham Beds in the Wincanton Borehole are explained by depositional thickening against a branch of the Mere Fault. Farther south, similar thickening occurred in a half-graben controlled by the Mere and Poyntington faults (Barton et al., 1993). There, clastic material may have been supplied by erosion of the Bridport Sands in local fault scarps.

After the Late Bathonian transgression, fault activity subsided, although thickness changes are associated with some faults, such as thinning across the Mere Fault in the Blackford area, or thickening into the Sherborne half-graben. Individual units are more persistent laterally and consistent lithologically than those lower in the sequence.

Lower and Middle Inferior Oolite

Lower and Middle Inferior Oolite strata are poorly represented in the district. In the north, the **Doulting Conglomerate** is the basal unit of the Inferior Oolite, belonging to the *subfurcatum* Zone, *banksi* Subzone (Parsons, 1975), and is equivalent to part of the Miller's Hill Beds of the Shaftesbury district. Richardson (1907) erroneously assigned the Doulting Conglomerate to the *garantiana* Zone. Although proved only in the Doulting area, it may form a basal conglomerate to the Inferior Oolite over much of the north of the district (Parsons, 1975). In the Doulting railway cutting (Figure 14, locality 2) [646 424], the unit comprises 0.4 m of fossiliferous, bioclastic limestones, with serpulid- and limonite-encrusted limestone and sandstone clasts (Parsons, 1975; Savage, 1977). The Garantiana Beds rest non-sequentially on the Doulting Conglomerate (Parsons *in* Cope et al, 1980b).

Farther south, fragmentary correlatives of the Corton Denham Beds, Miller's Hill Beds and Sherborne Building Stone (Table 6) occur in the Cole 'Syncline', near Bruton [667 336 to 684 348] (Figures 14 and 15). They comprise up to 2.3 m of sandy, sparsely glauconitic, very shelly biosparite, and include such units as the Pecten and Ammonite beds (Richardson, 1916; Parsons *in* Cope et al., 1980b). The deposits, bounded by and including many erosion surfaces, yield abundant faunas indicative of the *murchisonae* to *humphriesianum* zones (Parsons, 1979; Richardson, 1916) (Figure 15; Table 6). Exposures in the

Cole area (Figure 15, localities 10–16, 18) indicate many breaks in deposition. Brash (locality 10) [6716 3430], west of Bruton, close to the base of the Inferior Oolite yielded *Graphoceras* sp. and *Sonninia* sp. suggesting an interval from the uppermost *concavum* Zone to the lowest *discites* Zone (equivalent to Corton Denham Beds). In Lusty railway cutting (Figure 15, locality 11) [6817 3449], the Doulting Stone (Bed 1) and Hadspen Stone (Bed 2) rest on the eroded top of 1.4 m of limestones, sandy in part, and thin marls of the *discites* to *sauzei* zones (Beds 3–8). These rest disconformably on a 0.55 m-thick conglomeratic bed (Bed 9) of the *murchisonae* Zone, which overlies Bridport Sands. Parsons (in Cope et al., 1980b), following Richardson (1916), regarded the Pecten Bed (Beds 3–4) as ranging from the *laeviuscula* to *sauzei* zones; he placed beds 4a to 6 in the *laeviuscula* Zone, *ovalis* Subzone, and the 'Ammonite Bed' (Bed 8) in the base of that subzone down to the *concavum* Zone. In Lusty Quarry (locality 12) [6792 3442], the Garantiana Beds rest on ferruginous limestone (Bed 4) with a *humphriesianum* Zone, *blagdeni* Subzone fauna (Richardson, 1916). A new section [6721 3382] near Cole, which started about 6 m below the scarp edge, proved the equivalent of the Corton Denham Beds:

	Thickness m
CORTON DENHAM BEDS	
Micrite, peloidal, rubbly, hard, very shelly; less shelly in bottom 0.4 m. *Sphaeroidothyris* cf. *eudesi*, *Ctenostreon pectiniforme*, *Gresslya abducta*, *Inoperna plicatus* and *Graphoceras concavum* bedded in units of about 0.2 m; irregular erosion surface at base	1.2
BRIDPORT SANDS	
Sandstone, calcareous (0.2 m) and silty, fine-grained sand	0.7

At Sunny Hill Quarry (Figure 15) [672 337], Richardson (1916) recorded 3.4 m of Doulting Stone (Beds 1 and 2) and Hadspen Stone (Beds 3–9) resting non-sequentially on 2.3 m of strata belonging to the *discites* to *sauzei* zones (Beds 10–14). These older strata, the thickest correlatives of the Corton Denham and Miller's Hill beds in the district, include the Pecten Bed (Beds 10–12), ranging from the *sauzei* to *laeviuscula* zones. At Pitcombe (Figure 15, locality 14) [6740 3323], about 2 m of strata are overlain by the Hadspen Stone (Richardson, 1916); the Pecten Bed is missing and the Ammonite Bed (*discites* Zone) rests on the eroded top of limestones of the *murchisonae* Zone, which in turn rest on Bridport Sands. At Strutter's Hill (locality 15) [6748 3276], the sequence is thinner; 0.7 m of strata of the *discites* Zone rest on the eroded top of 0.4 m of limestones of the *murchisonae* Zone. Just to the south, at Mill Pitch (locality 17) [6772 3206], the Garantiana Beds rest on Bridport Sands (Richardson, 1916). The gamma-ray log of the Bruton Borehole [6896 3284] suggests a division into about 3.5 m of massive limestone at the top, probably the Doulting and Hadspen stones, overlying about 3.5 m of a more variable sequence of limestones and mudstones, possibly the Corton Denham and Miller's Hill beds. Some of the thin basal sequence may also occur in the south, just west of the district and north

of the Mere Fault (locality 24) [6505 2536], where 5 m of unexposed beds with a high gamma-ray source lie between the Hadspen Stone and Bridport Sands. This source may be equivalent to the phosphatic layer in the Miller's Hill Beds in the Purse Caundle Borehole (Barton et al., 1993). However, in nearby sections at Compton Pauncefoot (locality 23) [6520 2610], Windmill Farm (locality 25) [c. 644 236], Castle Hill Quarry, Shatwell (locality 19) [6611 2985], Hadspen (locality 18) [6550 3144], and Woolston (locality 22) [6606 2743], Hadspen Stone rests unconformably on Bridport Sands.

In the Wincanton Borehole [7116 2816], where the Inferior Oolite is apparently about 52 m thick, fossils indicative of the *concavum* and *murchisonae* zones, and equivalent to the Corton Denham Beds, occur about 10 m above the formation base (Ivimey-Cook, 1992a, unpublished report*). The overlying, fine-grained, grey limestone with irregular partings of dark grey silty clay are correlated tentatively with the Miller's Hill Beds. This suggests the presence of up to 33 m of Aalenian and Early Bajocian strata, and some 19 m of strata equivalent to the Hadspen Stone, Doulting Stone and Anabacia Limestone. Aalenian strata at the base suggest that the thickening was depositional, rather than the result of fault repetition. This interpretation is supported by the increase in dip with depth (from flat-lying at the top of the Inferior Oolite, to 45° at its base).

Upper Inferior Oolite

GARANTIANA BEDS AND HADSPEN STONE

The Garantiana Beds, also known as 'Ragstones', consist of up to 7 m of sparsely peloidal ragstone or limestone separated by thin marls; the base is commonly conglomeratic and upper surfaces of beds are commonly bored. In the south, the equivalent of the Garantiana Beds has been named the Hadspen Stone (Richardson, 1907; 1916) (Table 6). It consists of massively bedded, ferruginous, brown, sparry, detrital, very fossiliferous limestone, and correlates with the upper part of the Sherborne Building Stone farther south. It occurs in beds from 0.15 to over 1 m thick (Plate 2), with thin marly partings, and commonly with abundant brachiopods. The sand content is low (less that 0.2 per cent), which contrasts with the 'sandy' Sherborne Building Stone (Barton et al., 1993). A section [6753 3451] near Bruton shows 0.6 m of Hadspen Stone resting on Bridport Sands. It consists of peloidal, very shelly biosparite, conglomeratic at the base with abundant bivalves and clasts of Bridport Sands. In the Horsecombe Bottom Quarry (locality 18) (see below), the basal Hadspen Stone contains phosphatised fragments of ammonites from the Bridport Sands; in the Blackford area, the basal beds are also conglomeratic (Richardson, 1916).

The Garantiana Beds belong to the *garantiana* Zone, *acris* Subzone (Parsons in Cope et al., 1980b).

DOULTING STONE

The Doulting Stone (Richardson, 1907; 1916) is equivalent largely to the Rubbly Beds of the Shaftesbury district

Table 7 Correlation of the Great Oolite Group of the district with other areas.

Stage		Zone	Subzone	Sherborne area (after Cope et al., 1980b)	Shaftesbury district	Wincanton district	
CALLOVIAN	LOWER	Herveyi	Kamptus	Kellaways Clay	Kellaways Formation	Kellaways Formation	
			Terebratus	Upper Cornbrash c. 0.2–7.5 m	Upper Cornbrash 2.5–9 m	Upper Cornbrash 4–5 m	
			Keppleri				
BATHONIAN	UPPER	discus*	discus*	Lower Cornbrash c. 3.7 m	Lower Cornbrash 3–5 m	Lower Cornbrash 2–5 m	
			hollandi	Forest Marble 39 m	Forest Marble 37–55 m	Forest Marble 35–42 m	
		orbis*		Upper Fuller's Earth Clay 30 m	Frome Clay 45–65 m *(Frome Clay)*	Frome Clay 6–50 m	
		hodsoni*		Wattonensis Beds + 5 m	Wattonensis Beds 0–3 m	(Wattonensis Beds 3 m)	
				Middle Fuller's Earth Clay 9 m	Upper Fuller's Earth 2–21 m	Upper Fuller's Earth 0–9 m / Rugitela Beds 4–5 m	
	MIDDLE	morrisi*		*Fuller's Earth Rock* Ornithella Beds 0.6 m	*Fuller's Earth Rock* Ornithella Beds 0.6–3.4 m	Ornithella Beds 1.5–2.5 m	FULLER'S EARTH ROCK
				Linguifera Bed 0.6–1 m	Milborne Beds 6.1–7.3 m	Milborne Beds 6–7 m?	
		subcontractus*		Thornford Beds + 7.5 m			
		progracilis*		*Lower Fuller's Earth Clay* Acuminata Beds ?c. 1.5 m	*Lower Fuller's Earth* Acuminata Beds 5.6 m	Echinata Bed / Acuminata Beds ?	LOWER FULLER'S EARTH
	LOWER	tenuiplicatus		Clay	Bowden Way Beds 16–21 m	Clay	
					Hanover Wood Beds 19.1 m		
		zigzag*	yeovilensis*	Lenthay Bed 0.3 m / Clay 2 m	Knorri Beds 10.7 m / Lenthay Beds 6–8 m	Knorri Beds / Fullonicus Limestone 0–1	
			macrescens	Inferior Oolite	Inferior Oolite	Inferior Oolite	
			convergens				

(Right-hand vertical bracket: GREAT OOLITE GROUP)

* Zone/subzone proved in the Wincanton district

(Table 6). Around Doulting, the deposit comprises up to 9 m of white to yellow, massive, flaggy, pelbiosparite, commonly in fining-upwards packages, with common cross-laminated beds. Trough cross-laminated peloidal biosparite, 1 m thick, and in channels 2.5 m wide, is exposed in a quarry [6680 4046] on Small Down. These beds, between 0.5 and several metres thick (Plate 1), are separated by prominent, heavily bored hardgrounds. Burrows extend down to 0.2 m into the bed tops and are filled with poorly cemented iron-stained, peloidal sand with moulds of shells. Fossils, except the brachiopod *Acanthothiris spinosa*, are rare. The stone is yellow in contrast with the warm brown of the Hadspen Beds, and has been much quarried for building stone; for example, in the Doulting Quarries (Figure 15, locality 3) [between 649 435 and 653 436], just north of the district, where massive 'freestone' beds are well developed. Sections at Merehead [6925 4374 and 695 435], Cloford [7157 4430] and Holwell Quarries [725 449] just north of the district, expose Doulting Stone of the *parkinsoni* Zone (Savage, 1977), resting on Carboniferous Limestone. In contrast, nearby boreholes at East Cranmore [6838 4376], Merehead (locality 4) [6920 4363] and Cloford (locality 5) [7240 4340], just north of the district, proved, respectively, 12, 11 and 13.5 m of peloidal limestone on Ditcheat Clay of the Lias Group. A typical section in both the Doulting Stone and Hadspen Stone is that recorded by Richardson (1916) in Horsecombe Bottom Quarry [6550 3144] (locality 18), north-west of Hadspen House, where the following section can still be seen:

	Thickness m
DOULTING STONE	
Biosparite, peloidal, decalcified with shell debris	1.5
Biosparite, sparsely peloidal, bedding 0.2–0.45 m thick. Scattered whole terebratulids; coarse-ribbed bivalves, up to 0.15 m across, common 0.8 m above the base	1.8
Limestone, soft, marly, decalcified, ferruginous and rubbly, *Charltonithyris* sp., *Ptyctothyris*?, *Rhactorhynchia* cf. *hampenensis*, *R.* cf. *subtetrahedra*, *Sphaeroidothyris sphaeroidalis*, *Stiphrothyris tumida* and *Holectypus hemisphaericus*. Pronounced parting at base is probably an erosion surface	0.2
HADSPEN STONE	
Biosparite, peloidal to slightly peloidal, thin to thickly bedded in units, from top to bottom, 0.3, 0.3, 0.25, 0.35, 0.2, 0.2 and 0.45 m thick. Scattered bivalves and terebratulids. *Strenoceras* (*Garantiana*) cf. *garantiana*, 1 m from top, indicates the *garantiana* Zone, *acris* Subzone	3.0

Locally, as in a pit (locality 9) [6706 3560] on Creech Hill, there is a coral bed ('Upper Coral Bed') at the base of the Doulting Stone (Richardson, 1916, pit 32). However, it appears to be absent in pits (locality 8) [6691 3656 to 6690 3650] on the north side of Creech Hill.

In the south of the district, a section [6622 2403 to 6622 2407] in the Rubbly Beds and/or Doulting Stone near Golden Valley Farm (locality 26) showed:

	Thickness m
Soil and rubbly limestone head	0.50
?RUBBLY BEDS/DOULTING STONE	
Limestone, oolitic, sparry with irregular lenticular bedding, buff to cream-weathering; *Strigoceras truelli*	0.50
Marl parting, ochre to brown-weathering with small lenticular limestone bodies. Bivalves and fragmentary *Parkinsonia*	0.10
Limestone, moderately oolitic with some ferruginous ooliths, sparry, biomicritic. Irregular bedding, semi-nodular with a soft friable creamy buff-weathering matrix. Vertical burrows, 10 to 15 mm wide, are filled with marly matrix Limestones and matrix quite fossiliferous with *Cenoceras* cf. *inornatum*, *Procerites* sp. and *Acanthothiris spinosa* and *Sphaeroidothyris sphaeroidalis* 0.5 m from top	1.35

The Doulting Stone belongs to the *parkinsoni* Zone, *truelli* Subzone (Parsons in Cope et al., 1980b).

ANABACIA LIMESTONE

The uppermost division of the Inferior Oolite, the Anabacia Limestone, is recognised only around Doulting (locality 1) [6474 4320], just north-west of the district, where it is 3.4 m thick. The member, however, may form the uppermost part of the formation across most of the district. It correlates with the Crackment Limestones in the Shaftesbury district (Table 6) and comprises rubbly, peloidal, rarely ooidal, brown- or white-weathering limestone (Cope et al., 1980b). Richardson (1907) named the upper part of this unit the 'Rubbly Beds', but the beds are stratigraphically higher than the Rubbly Beds of the Shaftesbury district.

The Anabacia Limestone, which takes its name from the button coral *Anabacia*, ranges from the Upper Bajocian *parkinsoni* Zone, *bomfordi* Subzone, into the Lower Bathonian *zigzag* Zone, *macrescens* Subzone (Table 6).

GREAT OOLITE GROUP

The Great Oolite Group in the Bath type area (Penn and Wyatt, 1979) comprises four main units, the Fuller's Earth, Great Oolite, Forest Marble and Cornbrash formations. Marked facies changes occur southwards in the Group between Bath and Wincanton; in particular, the ooidal limestones of the Great Oolite Formation are replaced by mudstones of the Frome Clay (Figure 16) (Table 7).

There is little change in thickness in the formations of the group across the district, but farther south, the Fuller's Earth and Frome Clay are considerably thicker (Figure 16). Lithological and thickness changes in the higher formations, the Forest Marble and Cornbrash, are less significant.

Fuller's Earth (Formation)

The formation takes its name from the commercial seams of fuller's earth developed in the upper part of the sequence in the Bath area. The name has been in use

Figure 16 Diagrammatic cross-section through the Great Oolite Group between Bath and the south coast (data north of the district from Penn and Wyatt, 1979).

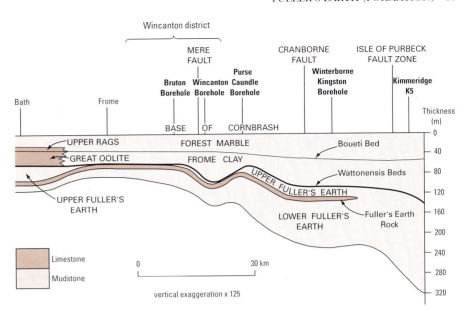

since the pioneering work of William Smith at the end of the eighteenth and beginning of the nineteenth century.

The Fuller's Earth Formation comprises two calcareous mudstone members, the Lower and Upper Fuller's Earth, separated by the Fuller's Earth Rock that consists mainly of bioclastic, micritic limestone with a few marl seams. Strata now regarded as the Frome Clay (Penn and Wyatt, 1979) were included in the upper part of the Upper Fuller's Earth during the previous survey.

The Fuller's Earth shows a southward thickening from the Mendip Hills, with most of the increase occurring in the Lower Fuller's Earth (Figures 16 and 17) and, to a lesser extent, in the Upper Fuller's Earth.

The formation ranges from the *zigzag* to the *hodsoni* Zone (Table 7).

LOWER FULLER'S EARTH (MEMBER)

The Lower Fuller's Earth has a narrow outcrop in the west of the district and usually occupies the face of a small, west-facing scarp capped by Fuller's Earth Rock.

At outcrop, the Lower Fuller's Earth thickens southwards from 3 m in the Whatley area, north of the district (Sylvester-Bradley and Hodson, 1957), to about 6 m on the northern margin of the sheet, to 16 m in the Norton Ferris Borehole, and to 31 m in the south. It may be up to 45 m thick at depth in the south-west of the district (Figure 17).

The member consists mainly of silty mudstone, but with some muddy limestone, mainly towards the top. In the Doulting railway cutting [650 425], just west of the district, the base of the Lower Fuller's Earth consists of 0.9 m of brown marls and thin cementstones, the **Fullonicus Limestone** (Torrens, 1969a; 1980) of the *zigzag* Zone, *yeovilensis* Subzone (Torrens, 1974); it is correlated with the Lenthay Limestone of the Shaftesbury district (Torrens, 1980), although it lacks the eponymous brachiopod *Sphaeroidothyris lenthayensis*.

Of the higher units recognised in the Shaftesbury district to the south (Barton et al., 1993; Bristow et al., 1995), only the **Knorri Beds** and **Acuminata Beds** are known with certainty in the present district, together with the **Echinata Bed**, which occurs above the Acuminata Bed in the Bath area (Penn and Wyatt, 1979) (Table 7).

The **Knorri Beds** (Barton et al., 1993), formerly known as the knorri Clays or *Ostrea knorri* Clays (Richardson, 1916), consist of grey, silty mudstones with common to abundant valves of the oyster *Catinula knorri*. In the dis-

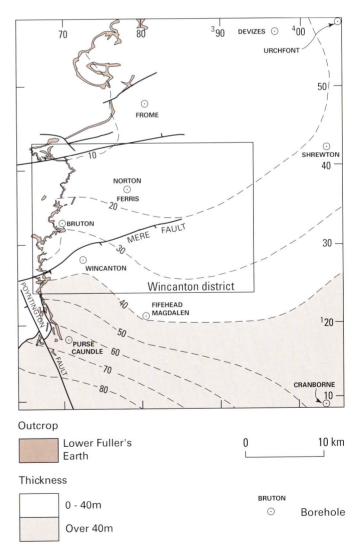

Figure 17 Isopachyte map of the Lower Fuller's Earth (contours at 10 m intervals).

trict, mudstones with *Catinula knorri* have been noted near Doulting (Richardson, 1909a) and on Creech Hill (Richardson, 1916). Common, argillaceous, shelly micrites (less than 0.2 m thick) towards the top of the Lower Fuller's Earth in the Bruton–Holton area may equate with the **Bowden Way Beds** of the Shaftesbury district (Bristow et al., 1995).

The **Acuminata Beds** (Barton et al., 1993), the *Ostrea acuminata* Beds of earlier authors, consist mainly of olive-grey, shell-detrital mudstones with the common oyster *Praeexogyra acuminata*. Shelly mudstones of this type were encountered in auger holes south-west [6544 4079] and north-west [6545 4241] of Chesterblade. Richardson (1909a) noted 'Ostrea acuminata' in the sides of Alham Lane [?6672 4133]. Ammonites (see below) were found in the Acuminata Beds by Torrens (1980) near Leighton [probably the pit at 6981 4343] and from west of Higher Alham Farm (exact locality unknown). An earthy limestone packed with *P. acuminata*, together with *Rhynchonelloidella wattonensis* and *Pholadomya*, occurs about 3 m from the top of the Lower Fuller's Earth in a section [6903 3503] in the River Alham near Bruton. Farther south, *P. acuminata*, *Rhynchonelloidella smithi*, *R. wattonensis* and *Catinula knorri* were found near Yarlington [6622 3088] a few metres above the base of the Lower Fuller's Earth. The co-occurrence of *R. wattonensis* and *P. acuminata* suggests a position high in the Lower Fuller's Earth, but *C. knorri* suggests a much lower position. Possibly, there is some mixing of the fauna due to slippage, although the interval between the Knorri and Acuminata Beds must be very small at this locality. At Shepton Montague, the Acuminata Beds were recognised from spoil [6894 3207] of grey silty clay with *Rhynchonelloidella smithi*, *R. wattonensis*, *Catinula* cf. *matisconensis*, *Pholadomya lirata* and *P. acuminata*. *Procerites* and *Wagnericeras* were found in the Acuminata Beds on the A303 near Dancing Cross [66 26] (Cope et al., 1980b). They are well developed in the Stowell (Taylor, 1990) and Purse Caundle boreholes (Barton et al., 1993).

An exposure [6675 2528], south of Maperton, of 0.5 m of grey clay with shelly muddy micrites with *Meleagrinella* cf. *echinata*, may be the **Echinata Bed**. This locality is probably the same as that recorded by Richardson (1909b) who noted 4.5 m of pale grey and 'whitish' clay with 'Pseudomonotis echinata' at the top of the Lower Fuller's Earth.

Diagnostic zone fossils are rare. Ammonites from the Fullonicus Limestone indicate the *zigzag* Zone, *yeovilensis* Subzone (Cope et al., 1980b; Torrens, 1974). This zone is also suggested by a *Zigzagiceras*? in the Knorri Beds, approximately 10 m above the base of the Fuller's Earth, in the Stowell Borehole south of the district. However, fauna from a section near Maperton [6719 2548] may indicate a slightly older age for the base of the member. There, *Parkinsonia (Gonolkites)*? and *Procerites*? which occur in a calcareous siltstone above grey silty clay, suggest the *zigzag* Zone (?*macrescens* Subzone). This fauna has also been found in the Crackment Limestones in the Shaftesbury district (Bristow et al., 1995), which also partly belongs to the *macrescens* Subzone (Penn, 1982; Barton et al., 1993). There is no ammonite evidence of the *tenuiplicatus* Zone, to which Torrens (1980) assigned the Knorri Beds. *Pro-*

cerites imitator and *Wagnericeras (Suspensites) suspensum* (Torrens, 1980) from the Acuminata Beds near Alham (see above) indicate the *progracilis* Zone.

FULLER'S EARTH ROCK (MEMBER)

The Fuller's Earth Rock forms a prominent scarp between Chesterblade in the north and Maperton in the south. Extensive dip slopes occur in the Chesterblade area and south of Bruton, but they are commonly covered by soliflucted Forest Marble debris.

The Fuller's Earth Rock varies from 4 to 7 m thick at outcrop; thicknesses proved in boreholes are 11.5 in Norton Ferris, 7.2 m in Bruton and 10.7 m in Wincanton boreholes. North of the district, the Fuller's Earth Rock is 8 m thick at Cloford [7240 4340] and near Whatley (Torrens, 1980).

The member comprises interbedded, shelly micrite with subordinate marly clay. The micrite is grey, weathering brown, 0.1 to 0.25 m thick, and commonly earthy. The clay is greyish brown, 0.02 to 0.1 m thick, with micritic limestone nodules. The micrite beds have very uneven tops and bases, deflecting laminations in the marls.

The Fuller's Earth Rock comprises three units; in ascending sequence, the **Milborne Beds, Ornithella Beds** and **Rugitela Beds**. The last two units are characterised respectively by an abundance of the brachiopods *Ornithella* and *Rugitela*.

A 4.06 m section [6882 3479] in Fuller's Earth Rock at Bruton railway station has been designated by Torrens (1974) as the European type section for the *morrisi* Zone. There, he recorded 0.93 m of the **Milborne Beds**, divided into a lower unit of rubbly limestones of the *subcontractus* Zone, overlain by 0.63 m of shelly, marly and rubbly limestones of the *morrisi* Zone, followed by 3 m of alternating thin beds of marl and limestone of the Ornithella Beds, and capped by 0.23 m of marls and limestones of the Rugitela Beds.

A good representatve section through the Fuller's Earth Rock, but with thinner Ornithella Beds, occurs at the Cliff Hill Quarry [6842 3229] (Freshney, 1994; Torrens, 1969b), but the lower part of the section is now obscured.

	Thickness m
RUGITELA BEDS	
Soil and rubbly limestone seen in tree roots.	
Acanthothiris powerstockensis, Rhynchonelloidella wattonensis, Rugitela bullata	
Gap	
ORNITHELLA BEDS	
Limestone	seen 0.15–0.23
Marl, *Rhynchonelloidella* very common	0.23
Limestone, rubbly	0.46
Marl, *Ornithella bathonica, Procerites* sp.,	0.30–0.36
Limestone, rubbly, *Ornithella bathonica, Procerites*	0.23–0.25
Polyzoa Marl	
Marl, soft, producing cut-back. Polyzoa (*Diastopora*), *O. bathonica, Wattonithyris* sp., *Procerites* spp.	0.18

MILBORNE BEDS

	Thickness m
Limestone in beds 0.15 to 0.38 m thick; corals in the lower part of the section	c.2.10

The Milborne Beds yielded *?Berbericeras* spp., *Lycetticeras* sp. and *Morrisceras morrisi* in the upper part (*morrisi* Zone), and *Tulites* sp. in the lower part (*subcontractus* Zone).

A road cutting [6678 2532] south-south-west of Maperton, the most complete exposure in the south of the district, showed:

	Thickness m
ORNITHELLA BEDS	
Biomicrite, hard with some scattered shells, rather, rubbly; *Ornithella pupa* in uppermost 0.15 m	1.50
MILBORNE BEDS?	
Biomicrite, muddy, slightly nodular, slightly shelly	1.40
Biomicrite, hard, buff, rather nodular. Some shells recrystallised	0.90
Siltstone, brown, calcareous with fine-grained shell debris	0.15
Biomicrite, nodular, fine-grained; no shells	0.10–0.15
Siltstone, calcareous	0.15

The ammonites *Oecotraustes (Paroecotraustes)* cf. *serrigerus* and *Tulites subcontractus*, indicative of the *subcontractus* Zone, and probably from the Milborne Beds, were obtained by Arkell (1951–1958).

The Milborne Beds must have been exposed in the 'Alham Lane Quarry' of Richardson (1909a) from where *Morrisceras morrisi*, *Tulites cadus* and *T. praeclarus* were collected. In a stream [6775 3071] north of Bratton Seymour, the Milborne Beds, consisting of 1 m of interbedded biomicrites and greyish brown marls, yielded *Kallirhynchia* cf. *platiloba*, *Rhynchonelloidella* cf. *mesoloba*, *R. smithi*, *R. wattonensis*, *Tubithyris powerstockensis* and *Lopha marshii*. From a nearby excavation [6826 3052], Mr H C Prudden obtained a varied fauna including cf. *Wagnericeras bathonicum* and *Belemnopsis fusiformis*.

Brash [6739 4174] of the **Ornithella Beds** from north of Alham yielded abundant *Ornithella bathonica*, together with *Montlivaltia* sp., *Procerites* cf. *quercinus* and *P.* sp. indicative of the *hodsoni* Zone. Southwards, brash [6689 2796, 6683 2802 and 6683 2805] from the Ornithella Beds yielded *Ornithella bathonica*, *?Ptychtothyris arkelli*, *Rhynchonelloidella smithi*, *Homomya gibbosa*, *Lopha marshii*, *Pholadomya lirata*, *Pseudotrapezium cordiforme*, *Sphaeriola oolithica* and *Procerites* sp.

In the north of the district, the Lodge Farm Borehole [6897 4133] penetrated 4.67 m of Fuller's Earth Rock (base not proved); the upper 4.41 m yielded *Rhynchonelloidella wattonensis*, *R. smithi*, *Acanthothiris spinosa* and *Rugitela bullata* and is regarded as **Rugitela Beds**. The Bratton Seymour Borehole [6763 2860] penetrated 4.36 m of Fuller's Earth Rock (base not proved), which is probably all Rugitela Beds.

The southern limit of the Rugitela Beds is uncertain; the most southerly record in the district is in the Bratton Seymour Borehole. It has been suggested (Sylvester-Bradley and Hodson, 1957; Torrens, 1980; Page, 1996) that the Rugitela Beds equate with the Wattonensis Beds which are developed south of the district. This was disputed by Penn and Wyatt (1979, p.50, fig. 14), who correlated the Wattonensis Beds of Dorset with the lower of the two Smithi limestones in the Bath-Frome area. The evidence presented here, whilst not conclusive, supports Penn and Wyatt's interpretation. If the Wattonensis Beds and the Rugitela Beds are the same, then the corollary is that where the Rugitela Beds cap the Fuller's Earth Rock, the Upper Fuller's Earth of the Wincanton and Shaftesbury districts (equivalent to the Middle Fuller's Earth Clay of Torrens (1980)) is absent. The absence of the Rugitela Beds at the top of the Fuller's Earth Rock at outcrop and in the Purse Caundle Borehole in the north of the Shaftesbury district, and the occurrence of 7 m of Upper Fuller's Earth separating the Wattonensis Beds from the Fuller's Earth Rock, might support this view. Although the presence of smectite is not diagnostic, the occurrence of high smectite in the Upper Fuller's Earth north of the Mendips (Penn and Wyatt, 1979) and in the Purse Caundle Borehole (Barton et al., 1993), suggests that in this district, the 8 m of smectite-rich mudstones (Mitchell and Murphy, 1995) below presumed Wattonensis Beds and above the Rugitela Beds in the Lodge Farm and Bratton Seymour boreholes, are also Upper Fuller's Earth. The presumed Wattonensis Beds in these boreholes correlate geophysically with the Wattonensis Beds in the Purse Caundle Borehole.

?Berbericeras spp., *Lycetticeras* spp., *Morrisceras* and *Holzbergia* found in the upper part of the Milborne Beds, and *Tulites* and *Trolliceras* in the lower part indicate, respectively, the *morrisi* and *subcontractus* zones. The Ornithella and Rugitela Beds fall in the lower part of the *hodsoni* Zone (Table 7; Torrens 1980).

Ostracods from the Lodge Farm and Bratton Seymour boreholes indicate that the *polonica/confossa* ostracod biozonal boundary falls in the Rugitela Beds (Wilkinson, 1994a, unpublished report*; 1994b, unpublished report*). These two zones approximate respectively to the *progracilis–morrisi* and *hodsoni* ammonite zones.

UPPER FULLER'S EARTH (MEMBER)

The Upper Fuller's Earth is mapped only in the south as a projected continuation of the outcrop in the Shaftesbury district where the Wattonensis Beds are well developed. Elsewhere in the district, the Upper Fuller's Earth is included with the Frome Clay. Geophysical logs suggest that the Upper Fuller's Earth is present in the Lodge Farm (8.4 m thick), Bratton Seymour (8 m thick) and Norton Ferris (9 m thick) boreholes (Figure 18). The member thickens to about 28 m in the Bath area (Penn and Wyatt, 1979).

The Upper Fuller's Earth consists mainly of pale olive-grey, silty clay with a few calcareous siltstones or silty muddy micrites. Darker olive-grey clay also occurs; this is patchily lignitic, with shells which are possibly kerogen-coated. Lamination is picked out by colour banding, but burrowing is also common. Shell-detrital material is relatively abundant, as well as intact bivalves, ammonites and brachiopods. The clay minerals are dominantly illite and

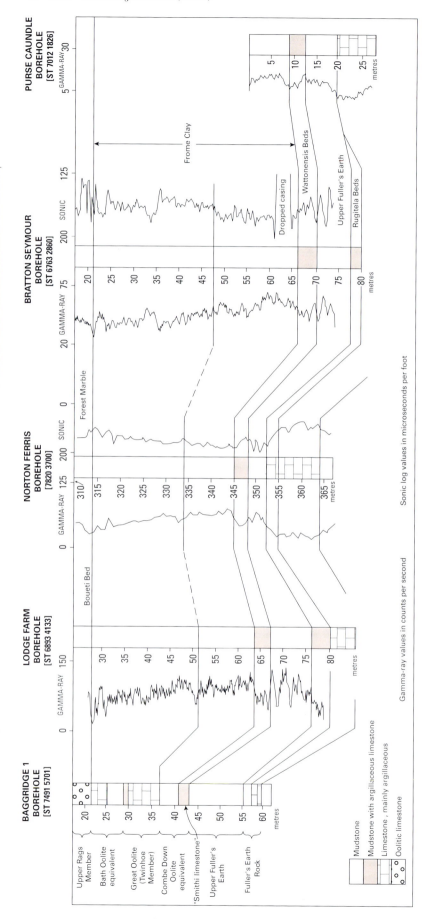

Figure 18   Correlation of part of the Fuller's Earth and the Frome Clay across the district.

kaolinite, but smectite constitutes up to 26 per cent by weight in the Bratton Seymour Borehole, and up to 15 per cent in the Lodge Farm Borehole (Mitchell and Murphy, 1995). These abundances compare with up to 35 per cent in the Purse Caundle Borehole (Barton et al., 1993).

The fauna from the Upper Fuller's Earth includes the brachiopods *Kallirhynchia expansa*, *Rhynchonelloidella smithi* and *R. wattonensis*, bivalves such as *Praeexogyra hebridica subrugulosa* and the ammonites *Oecotraustes* and *Oxycerites*. In the Lodge Farm Borehole, ostracods from between 66.31–66.34 m indicate the lowest part of the *blakeana* ostracod Zone (= high *hodsoni* Zone). A sample some 7 m lower yielded ostracods indicative of the *polonica* ostracod Zone (also falling in the *hodsoni* Zone) (Wilkinson, 1994a, unpublished report*). Sheppard (1981) placed the *blakeana/polonica* zonal boundary at, or just above, the base of the Wattonensis Beds (see below).

Frome Clay (Formation)

The mainly mudstone sequence between the Upper Fuller's Earth and the Forest Marble in north Somerset was named the Frome Clay by Penn and Wyatt (1979). Southwards from the type area, and including the whole of this district, the Frome Clay was formerly included in the Upper Fuller's Earth.

The formation crops out in the west of the district and forms gently sloping ground at the foot of the Forest Marble escarpment. Debris from the latter formation, possibly from mudflows, obscures much of the outcrop. The Frome Clay is about 32 m thick in the north, thickening to over 70 m in the south (Figure 19).

The base of the formation is taken at the base of the Wattonensis Beds (Penn and Wyatt, 1979), a thin unit of calcareous mudstones and argillaceous limestones. However, in this district, the Wattonensis Beds have fewer limestone beds than farther south and have proved unmappable. Consequently, the Frome Clay includes the thin Upper Fuller's Earth.

Above the Wattonensis Beds, the Frome Clay consists of olive-grey, calcareous, silty clay with some slightly carbonaceous, brown clay. At the top, pale grey, very silty clay and silt show up well in ploughed fields, stream sections (for example north-west [7030 2938] of Wincanton, south-east of Bruton [7038 3424 and 7036 3465], and south-west of Wanstrow [7065 4085]) and ditches.

The most complete sections in the district are the Lodge Farm and Bratton Seymour boreholes. Comparison of the gamma-ray log of the Lodge Farm Borehole with that of the Purse Caundle borehole (Figure 18), suggests that micritic limestone beds between 64.2 m and 67.5 m depth are the Wattonensis Beds. They yielded a fauna which includes *Discinisca* sp., *Kallirhynchia?*, *Rhynchonelloidella smithi*, *Bositra buchii*, *Camptonectes laminatus*, *Grammatodon bathonicus* and *Modiolus imbricatus* and the ammonite *Oecotraustes (Paroecotraustes)* aff. *serrigerus*. The Frome Clay above the Wattonensis Beds comprises 40.6 m of pale to medium olive-greys silty to fine-grained sandy mudstone, with darker olive-grey, less silty bituminous mudstone, in units up to a metre thick, becoming more common

towards the base. Because of extensive bioturbation, the mudstones are generally structureless and poorly fossiliferous; thin beds of micrite are fairly common. Shelly material is rare in the upper part, but becomes very common towards the base in the darker mudstones.

In the Bratton Seymour Borehole, the Frome Clay is 48.7 m thick. The Wattonensis Beds, recognised on geophysical evidence only, are overlain by 12.7 m of dark greenish grey, silty, calcareous clay. Ammonites, mainly oppeliids, including *Oxycerites* and *Oecotraustes?*, are common. The highest part of the Frome Clay comprises 36 m of olive-grey, calcareous, silty to very silty mudstone with some very sandy mudstone. Thin, more calcareous layers are fairly common, as is colour banding, but this is commonly disrupted by bioturbation. Branching, sub-horizontal and vertical, burrows occur. Sparse, fine-grained lignitic debris is common. The fauna in this part is mostly bivalves such as *Bositra buchii*, *Parainoceramus*, *Protocardia* and *Thracia?*. A few rhynchonellids, including *R. smithi*,

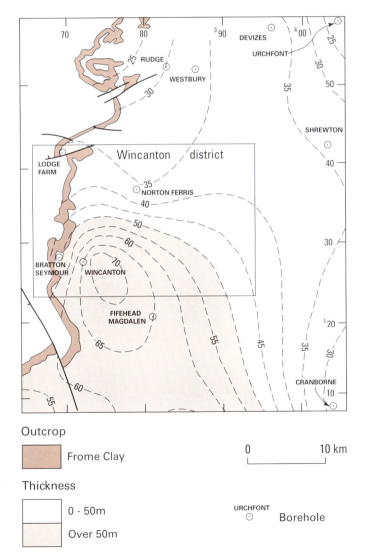

Outcrop

Frome Clay

Thickness

0 - 50m

Over 50m

0   10 km

URCHFONT
⊙ Borehole

Figure 19 Isopachyte map of the Frome Clay (contours at 5 m intervals).

and oppeliid ammonites, including *Oxycerites*, occur low in this section.

Macrofaunas from these boreholes are long-ranging, zonally undiagnostic taxa; the poorly known oppeliid ammonites may repay further investigation. The ostracods show that the Frome Clay belongs to the *blakeana* ostracod zone, equivalent to the late *hodsoni* and *orbis* [formerly *aspidoides*] zones. Outside the district, rare ammonites indicate that the Wattonensis Beds fall in the *hodsoni* Zone (Torrens, 1980), and the rest of the Frome Clay in the late *hodsoni* and *orbis* zones (Table 7).

East and north of the district, the Frome Clay passes from a mainly argillaceous formation to one of shelf limestone. The most northerly, fully argillaceous, occurrence of Frome Clay is in the Frome Borehole (Penn and Wyatt, 1979), where the lower part, equivalent to the Combe Down Oolite, comprises three cycles; the lower part of each cycle consists of silty calcareous mudstone, and the upper part of black shaly mudstone. Similar dark shaly beds with paler calcareous mudstones occur in the Lodge Farm and Bratton Seymour boreholes, but evidence of cyclicity is best seen in the geophysical logs, especially in the latter borehole. These logs also show three cycles, in which there is a steady diminution in the gamma-ray count, with an attendant rise in sonic velocity (Figure 18).

Palaeogeography of the Fuller's Earth and Frome Clay

Regional thickness variations in the Fuller's Earth and Frome Clay sequences suggest that the rates of subsidence in the Wessex Basin (including the Mere Basin) south of the Mere Fault increased steadily during the Bathonian and led to the development of a basinal mudstone facies. The Fuller's Earth Rock was probably deposited in shallower water than that of the Lower Fuller's Earth and Frome Clay.

Clay-mineral determinations from the Lodge Farm, Bratton Seymour and Purse Caundle boreholes (Barton et al., 1993; Mitchell and Murphy, 1995) show a low kaolinite to illite ratio throughout the Fuller's Earth and Frome Clay. Smectite is absent in the upper part of the Lower Fuller's Earth and Fuller's Earth Rock, becomes significant in the Upper Fuller's Earth and shows a marked decrease, or is absent, in the Frome Clay. Similar trends occur in the Winterborne Kingston Borehole (Knox, 1982). Ostracods in the lower part of the Upper Fuller's Earth in the Lodge Farm Borehole are less diverse than that from the upper part and may suggest that the lower part was deposited in more brackish conditions. The Upper Fuller's Earth thickens northwards towards Bath and, unlike the Frome Clay which passes into shelf limestone, continues in mudstone facies. Foraminifera are rare and of low diversity in the Frome Clay of the Bratton Seymour Borehole. This is probably due to a low-salinity palaeoenvironment (the majority of the ostracods are tolerant of wide salinity variations).

The transition southwards and westwards from the shelf limestones of the Great Oolite of the Bath–Hampshire area, to mudstones of the Frome Clay in this district, does not seem to be a progression into a deeper marine basin. The macrofauna, although sparse, is generally marine

and includes both benthic and planktonic taxa. However, brackish-water-tolerant ostracods at some levels suggest that the basin at times may have had restricted marine access, and this, combined with a significant input of fresh water, reduced salinity. The more varied fauna in the Wattonensis Beds and overlying 10 m of strata indicates a better connection to fully marine conditions. The trend of the facies change between limestone and mudstone, and the trend of the thinning of the limestones are both aligned north-west, parallel to one of the structural elements in the region.

Forest Marble (Formation)

The Forest Marble is a widespread formation which takes its name from the Old Forest of Wychwood in Oxfordshire, where limestone which could be polished was once quarried. The name was first adopted for geological purposes by William Smith in 1799.

The north-trending crop of the Forest Marble is offset by the Warminster and Mere faults (Figure 20). The western edge of the outcrop is commonly a prominent scarp.

The Forest Marble has the most uniform thickness of the Jurassic formations in the district, and is typically between 35 to 40 m at or near outcrop (Figure 20). The Norton Ferris Borehole [7820 3700] proved 41.76 m. The 31.4 m of marls and limestone proved in the Wincanton Borehole [7116 2816] (Edwards and Pringle, 1926; Ivimey-Cook, 1992a, unpublished report*) may be a minimum figure only as part of the sequence may be faulted out. At Bayford, a borehole [7287 2884] indicates a minimum thickness of 38.8 m of Forest Marble. North of the district, the Forest Marble thins north-eastwards to about 30 m, but increases towards the south and south-west to over 45 m (Figure 20).

The base of the formation is taken at the base of the Boueti Bed, a yellowish grey, muddy, shelly micrite, up to 0.3 m thick, that overlies the Frome Clay non-sequentially. The fauna of this basal bed consists of the diagnostic and eponymous brachiopod *Goniorhynchia boueti* together with *Digonella digonoides*, *D. digona* and *Avonothyris*, serpulids, corals and bivalves, including *Praeexogyra hebridica*. North of the district, the Boueti Bed is absent and its place at the base of the Forest Marble is taken by a locally developed shell bed with a **'Bradford Clay** fauna' that also includes *Digonella digona* (Penn and Wyatt, 1979). In this district, the co-occurrence of *D. digona* and *G. boueti* in the Boueti Bed near Redlynch [6967 3324] (see below) supports the correlation of these two beds between south Dorset and the Mendips.

A higher shell bed, also with a Bradford Clay fauna, is present at the top of the **Upper Rags** in the Bath area (Penn and Wyatt, 1979); it lies some 7 m above the base of the formation. In south Dorset, a comparable shell bed, the **Digona Bed**, also with *D. digona*, known as far north as the Winterborne Kingston Borehole (Penn, 1982), is correlated with the upper Bradford Clay shell bed (Penn and Wyatt, 1979; Penn, 1982), although there is no proved occurrence of the Digona Bed in the intervening, poorly exposed, ground.

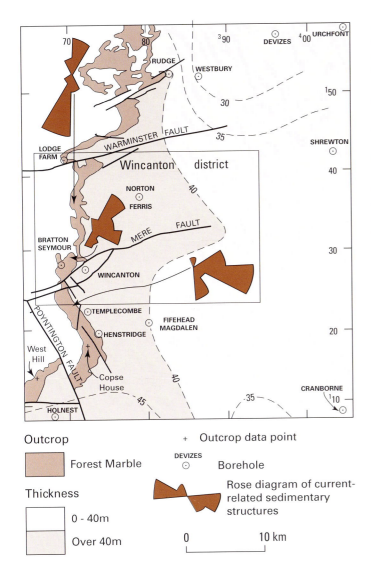

Figure 20   Isopachyte map (contours at 5 m intervals) and rose diagrams showing orientation of sedimentary structures of the Forest Marble.

The top of the Forest Marble is taken at the base of thickly bedded, micritic limestones of the Cornbrash which commonly form a featureless feather edge on the Forest Marble dip slope. West of Charlton Musgrove, lithified units within the Forest Marble are scarce and the base of the Cornbrash defines the scarp.

The formation name suggests a predominance of limestone, but the sequence is highly variable and the main lithology is a greenish grey, weathering brownish grey, sandy, locally marly mudstone. Sandy mudstone predominates in the lower part of the formation, particularly in the south of the district. Orange, calcareous sandstone is common throughout the sequence as thin wisps, lenses, laminae and beds, some with ripple cross-lamination. Thick bodies of reddish orange sand, resembling the **Hinton Sands** of the upper Forest Marble north of the Mendips, occur in places. The sandstones occur as lami-

nated or cross-bedded, calcareous units with trails of the trace fossil *Gyrochorte comosa* on the surfaces of thinly bedded sandstones. Laterally impersistent limestones, up to 6 m thick, such as at Knowle Rock Farm [7032 3119] (Plate 6), south-west of Stoney Stoke, occur throughout the sequence, but are more typical of the upper part. The limestones vary from cross-bedded, sparsely ooidal, oyster-rich limestones, commonly with mud clasts and abundant lignite debris in their basal part, to massively bedded, very shelly sparites.

Cemented units vary both vertically and laterally, from calcareous sandstone to limestone. Tabular cross-stratification is relatively common in thick lithified units, with cross sets defined by the variation in abundance of sand, shells and ooidal micrite. The sets range in thicknesses from 0.1 to 1 m, typically with a lateral extent of several metres. The laminae are commonly planar inclined at about 30°, and in part, become tangential to the base of the set. Climbing ripples are locally preserved.

Elongate erosional scours, typical of trough-cross stratification, also occur locally and consist of a succession of gently plunging curved laminae. Trough cross-stratification sets are less than 1 m thick and mainly parallel to the direction of plunge. Herringbone sets, in which dip reversal occurs in narrow packages of 50 to 200 mm, occur rarely with trough cross-stratification. Irregular subhorizontal surfaces, across which cross-laminae are interrupted (reactivation surfaces), are present in sandstones and sandy limestones. Limestones are packed with bivalves which lie parallel with the bedding and clearly have been winnowed.

A section [7068 2899] below Wincanton hospital, in the upper part of the formation, shows its variable nature:

	Thickness m
Clay, sandy to extremely sandy with common 2–3 mm sandstone or sand lenses; some thicker (50 mm) lensoid fine-grained sandstones	2.50
Sand, clayey, fine-grained, bioturbated with greenish grey clay clasts	0.20
Clay, sandy, greenish grey interlaminated with orange fine-grained sand in laminae (2–3 mm thick); some thin sandstone lenses	1.80
Biosparite, yellow and grey, slightly ooidal, shelly Some lignite debris.	0.10
Clay, sandy, greenish grey with sand laminae	0.40
Sand, fine-grained, orange with laminae of greenish grey clay. Some sand-filled burrows.	0.70
Sandstone, hard, calcareous, fine-grained	0.15
Clay, greenish grey with sand lenses and laminae (2–5mm)	0.10
Sandstone, very hard, calcareous, fine-grained. Ripple marks and clay drapes.	0:20
Clay, greenish grey, interlaminated with fine-grained sand.	0.05
Sandstone, very hard, brownish grey.	0.10
Sand, fine-grained, orange with greenish grey clay laminae.	0.10
Sand, orange, fine-grained	0.30

A typical section [6967 3324] in the Boueti Bed near Redlynch shows:

Plate 6 Carious, cross-bedded Forest Marble limestone, Knowle Rock Farm. Hammer length 30 cm. (A15553)

	Thickness m
FOREST MARBLE	
Biosparite, ooidal to oosparite in units of 0.04–0.18 m with a 0.02 m-thick marl seam about half way up. Some cross-bedding of 0.1 m scale. Mud clasts common. Lowest 0.1 m partly decomposed to ferruginous clay	0.80
Clay, greenish grey	0.02
Earthy layer, highly ferruginous.	0.01
Boueti Bed: silt to sandy silt, cream coloured with some nodular cementation; *Acanthothyris?, Digonella digona, G. boueti, Rhactorhynchia argillacea* and small bivalves	0.12
FROME CLAY	seen

A section through the Boueti Bed, with *D. digona*, but without the *G. boueti*, was seen in a stream [7036 3465 and 7038 3424] east of Bruton. The Boueti Bed was proved in the Lodge Farm, Bratton Seymour and Wincanton boreholes; it was also recognised by its distinctive high sonic velocity and low gamma-ray signature in the Norton Ferris Borehole (Figure 18). The Boueti Bed has been noted at the following localities: near Upton Noble [7091 4062], where it consists of 1 m of white-weathering, calcareous mudstone with *Avonothyris* sp. and *G. boueti*; temporary exposure [7017 3531] near Sheephouse Farm, north-east of Bruton (information from Professor D T Donovan, April 1996); as brash with *G. boueti* near Cranmore [682 417] (Torrens, 1980), and near Bratton Seymour [6763 2888, 6760 2861, 6738 2793, 677 292] (Freshney, 1994; Torrens, 1980). It persists southwards, through Dorset, to the coast (Cope et al., 1980b). Locally in the district, the Boueti Bed is absent, as in the road cutting [6962 3762 to 6967 3757] 1 km north-east of Henley Grove. Its occurrence in the Lodge Farm Borehole is its northernmost record.

Sections in the mudstones are generally rare, but there are exposures [7077 3576, 7104 3704, 7088 3568 and 7092 3567], up to 6 m high, in river cliffs along the River Brue and its tributaries west of South Brewham, where mudstones with interbedded sandstones occur.

A representative section in sandstones occurs in a stream [6982 2802] near Hook Valley Farm. In descending order, the sequence consists of 0.35 m of fine-grained, calcareous, flaggy sandstones: 0.3 m of ooidal sandstone to sandy oolite: 0.2 m of yellowish grey clay: 0.2 m of ooidal sandstone, resting on grey clay.

Limestone, sandstone and sand bodies are extensive south of Wanstrow, where they form very undulating topography with well-defined dip slopes, and around Bratton Seymour. A representative section occurs in a quarry [6887 2593] in North Cheriton:

	Thickness m
Biosparite, ooidal, thinly bedded (10–20 mm), thicker bedded towards base. Bivalves abundant	1.90
Biosparite, ooidal, thin- to medium-bedded up to 100–150 mm; some cross-bedding	3.00
Clay, grey, and interlaminated, fine-grained sand, lignitic	0.02
Biosparite, ooidal, thin to medium-bedded	0.90

South of North Cheriton, an excavation [6907 2535] exposes the contact between the Forest Marble and Cornbrash:

	Thickness m
CORNBRASH	
Biomicrite, rubbly, with terebratulids and bivalves	0.2
FOREST MARBLE	
Clay, extremely sandy, orange-brown and yellow, to clayey silty sand; sand-free in basal 0.4 m	1.1
Biosparite, hard, shelly	0.2
Clay, grey, sticky, with a few thin (20–30 mm) shelly biosparite layers. Thin laminated sandstone near the top	1.1
Biosparite, very shelly, medium bedded (up to 0.3 m)	0.6

The Forest Marble spans the Upper Bathonian *orbis* and *discus* zones (Dietl, 1982; Torrens, 1980) (Table 7), but ammonites, on which the standard zonation is based, are rare. Ostracods from the Lodge Farm [6897 4133] and Bratton Seymour [6723 2860] boreholes support and refine the ammonite zonation. The upper strata of the Frome Clay (highest dated sample 8 m below formation top) contain faunas characteristic of the *blakeana* ostracod Zone (equivalent to the upper part of the *orbis* Zone). The lowest occurrence of *Micropneumatocythere falcata*, the inception of which is at or near the base of both the *falcata* ostracod Zone and *discus* ammonite Zone, was just above the Boueti Bed in the Lodge Farm Borehole [6897 4133], as in Dorset (Sheppard, 1981).

LODGE FARM AND BRATTON SEYMOUR BOREHOLES

Both boreholes were sited on the Forest Marble and proved the lowest part of the formation (Figure 18). The Lodge Farm Borehole [6897 4133] proved an incomplete 28 m of interbedded pale olive-grey, silty to sandy mudstone with beds of sand and shelly sparite up to several metres thick. Calcareous sandstone occurs as millimetre-thick lenses. The basal part of the formation consists of 2 m of shelly sparite with muddy patches overlying in turn, a thin bed of mudstone, and a thin basal conglomerate with micritic pebbles in a bioturbated, shelly sparite matrix, possibly equivalent to the Boueti Bed. Lithified beds are common in the upper part of the sequence; thin, lensoid beds of more or less cemented sandstone and up to 1 m-thick beds of sand or shelly sparite are present. The limestone is medium to coarse grained, has abundant broken shell fragments, and is locally ooidal. Sandstones are grey, calcareous, fine grained, commonly with load casts and abundant lignitic debris; the beds are associated with sandy mudstone with lenses and laminae of sand.

The Bratton Seymour Borehole [6763 2860] proved the lowest 21.33 m of Forest Marble. Above the Boueti Bed, the formation consists of brecciated mudstone which is bioturbated, pale greenish grey, silty to sandy and finely micaceous. In places, the mudstones are structureless and homogeneous. Sparsely developed throughout are silt partings and poorly preserved laminae, together with wisps and lenses, 5 to 10 mm thick, of silt, sand, calcareous sandstone and sandy limestone. Sand-filled horizontal burrows, up to 20 mm diameter, abound in the basal 5 m. At 15.67 m above formation base, there is a fine-grained, greenish grey, clayey micrite (0.36 m thick) with horizontal and inclined burrows (5 mm diameter) and filled with medium grey limestone or silt. The micrite also shows silty lamination, with a 10 to 20 mm spacing, and silt lenses up to 8 mm thick; the silts are laminated with scours on their top surface; shells abound in the basal 20 mm of the micrite.

PALAEOGEOGRAPHY

The sedimentary structures suggest deposition in shallow water, in which flow varied in direction and strength. Tabular cross-stratification is indicative of relatively low flow strength, compared with trough sets which indicate more vigorous flow and are typical of shallow-water deposition. Climbing ripples suggest the operation of reverse eddies; reactivation surfaces and herringbone structures may indicate changes in direction of flow or tidal reversals.

Orientation data (Figure 20) show that inferred flow directions fall mainly in a north-east to south-east quadrant, similar to those in the north of the Shaftesbury district (Bristow et al., 1995). In the north, tabular beds dip southwards, but north-inclined laminations, particularly in herringbone sets, form a subsidiary component. Trough axes are generally aligned north-north-east, but in a pit [7087 4071] south of Wanstrow, a channel up to 7 m wide and 1 m deep trends east–west and is infilled with 4 m of massive to thick-bedded, shelly, peloidal sparite. An exposure in a stream [7031 3412] north of Redlynch, shows mega-ripples in sandstone that trend 225° and have an amplitude of 0.1 m and wavelength of 0.8 m. Overall, the observations suggest a dominantly south- or west-directed palaeocurrent, with periodic reciprocal flow (Figure 20).

The restricted form of lithified units, particularly the shell-detrital limestones, suggests that they formed as shoals on a muddy sea floor. Mud clasts and large wood fragments in the limestones suggest deposition during storm-related events in which oyster fragments and ooids were introduced.

The oyster-dominated and trace-fossil-dominated faunas, the low-diversity foraminiferal assemblage, the abundance of land-derived plant debris and absence of pronounced bioturbation is consistent with a salinity less than fully marine, as suggested by Hallam (1970).

Cornbrash (Formation)

The term Cornbrash is an historical name that originated in Wiltshire and is derived from the stony brash developed on the formation and on which corn is commonly grown.

The formation is a thin limestone unit, typically with a wide crop on well-defined dip slopes, such as between Upton Noble and Charlton Musgrove. The outcrop is offset by the Warminster, Witham Friary and Mere faults (Figure 21).

The Cornbrash thickens southward from about 2 m in the Mendip area to more than 10 m south and east of the district (Figure 21). Some 4.26 m were proved in the Norton Ferris Borehole. The formation is reputedly (Freshney, 1994) 6 m thick in a pit at Suddon Farm [698 289], west of Wincanton, and 4.6 m thick in a borehole [6986 2772] to the south. In the Combe Throop Borehole [7260 2350], just south of the district, the formation is 6.87 m thick, comprising 3.74 m and 3.13 m of Lower (mainly Obovata Beds, see below) and Upper Cornbrash respectively. The 8.23 m recorded in the Wincanton Borehole (Edwards and Pringle, 1926; Freshney, 1995a; Ivimey-Cook, 1992a, unpublished report*) is assumed to be incorrect, as the nearby Bayford Borehole [7287 2884] proved only 5.1 m of Cornbrash. The basal 3.43 m of strata in the Gillingham Borehole (Whitaker and Edwards, 1926) are here all attributed to the Cornbrash and give only a minimum thickness for the formation (compare with Bristow et al., 1995, fig. 30).

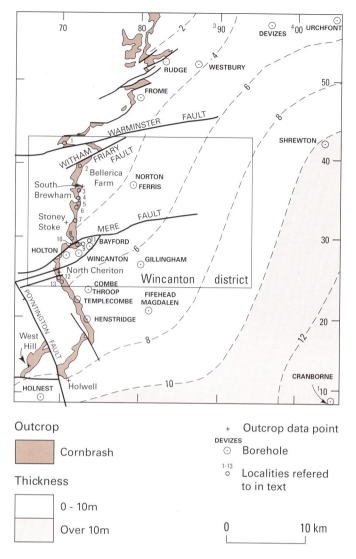

Outcrop

Cornbrash

Thickness

0 - 10m

Over 10m

+ Outcrop data point

DEVIZES
⊙ Borehole

1-13
° Localities refered
to in text

0   10 km

Figure 21 Isopachyte map for the Cornbrash (contours at 2 m intervals).

The formation is divided traditionally, on lithology and fauna, into Lower and Upper Cornbrash which approximate respectively to the Berry and Fleet members of Page (1989).

Ammonites in the Lower Cornbrash, such as *Clydoniceras discus* and *Homoeoplanulites homoeomorphus* from Wincanton [7132 2877] (Page, 1988), indicate the Upper Bathonian *discus* Zone and Subzone; ammonites in the Upper Cornbrash indicate the Lower Callovian Herveyi (formerly *macrocephalus*) Zone (Cope et al., 1980b; Page, 1989). Although ammonites provide the standard stratigraphical control, brachiopods, which predominate, have also been used (Buckman, 1927; Douglas and Arkell, 1928; 1932). Four brachiopod biozones, named after *Cererithyris intermedia* and *Obovothyris obovata* in the Lower Cornbrash, and *Microthyridina siddingtonensis* and *M. lagenalis* in the Upper Cornbrash, are recognised. Arkell (1933, p.334)

showed that the distribution of *C. intermedia* and *O. obovata* is probably facies controlled.

LOWER CORNBRASH (MEMBER)

The Lower Cornbrash comprises pale cream, sparsely ooidal and peloidal biomicrite, locally very shelly or with thin, shelly mudstone partings and pseudonodular appearance. Thick cross-bedded limestones occur locally.

Four principal units were recognised by Douglas and Arkell (1928) in north Somerset and Wiltshire (Figure 22). The lowest unit, the **Intermedia Bed,** consists of micritic limestone with abundant *Cererithyris intermedia*. A median unit consists of either the **Corston Beds** or its lateral equivalent, the **Obovata Beds.** The Corston Beds comprise shell-detrital and sparsely fossiliferous limestones, typically cross-bedded in sets up to 3 m thick. The Obovata Beds comprises argillaceous, pseudonodular limestone with *Obovothyris obovata*. The uppermost unit, the **Astarte–Trigonia Bed,** comprises a richly fossiliferous micrite characterized by a bivalve assemblage of *Astarte* [=*Neocrassina*] *hilpertonensis* and trigoniids. Although most units can be recognised in sections, they are too thin to map separately. Facies and fauna can vary rapidly, however, and more than one non-sequence may occur; locally, some of the above-named units may be missing.

The **Intermedia Bed** is laterally persistent over much of Dorset and south Somerset, its presence is indicated by brash near Redlynch which yielded *Cererithyris intermedia* and *Pholadomya lirata* [7186 3380] (Figure 21, locality 6a) and *C. intermedia*, *Kallirhynchia globularis* and *Pleuromya uniformis* [7170 3387] (locality 6b). A section south of North Cheriton [6907 2535] (locality 14) showed 0.2 to 0.3 m of rubbly biomicrite with *C. intermedia*, resting on Forest Marble. However, the bed is locally absent, for example in a pit [probably the one at 721 394] (locality 2) at Bellerica Farm, where up to 1.8 m of flaggy, sparsely fossiliferous limestones (Corston Beds) rest on ooidal marl of the Forest Marble, at Wincanton Racecourse (locality 8) (Douglas and Arkell, 1928), and just to the south in the Combe Throop Borehole (Bristow et al., 1995).

The **Corston Beds** are particularly well developed north of Frome, while their lateral equivalents, the **Obovata Beds,** occur principally south of the Mere Fault; the two facies probably interdigitate across much of the district. The section at Cards Farm (Figure 21, locality 4), South Brewham, showed flaggy, cross-bedded, unfossiliferous Corston Beds, overlain by 0.15 m of marly clay with *Ornithella obovata* (Douglas and Arkell, 1928). The Corston Beds (at least 2 m thick) were exposed in the A303 cuttings (locality 11) at Bayford (Bristow et al., 1992), and near South Cheriton (locality 13), where it consists of 1.8 m of sparsely fossiliferous limestone, blue-hearted at the base, massive and white above, and rubbly and nodular at the top.

The **Astarte–Trigonia Bed** is well exposed at South Brewham (Figure 21, locality 3) and South Cheriton (locality 15), where it is 0.4 m thick and yielded *Clydoniceras* sp. It is absent near Wincanton (Douglas and Arkell, 1928), Bayford (locality 11) (Bristow et al., 1992) and in the Combe Throop Borehole.

N   Wincanton district   S

FROME [7895 4720]

2 BELLERICA FARM, UPTON NOBLE [?721 394]

3 SOUTH BREWHAM [7165 3616]

4 CARDS FARM, SOUTH BREWHAM [722 347]

7 STONEY STOKE [7415 3190]

NORTON FERRIS BOREHOLE [7820 3700]

10 WINCANTON [7115 2904]

12 BAYFORD BOREHOLE [7287 28842]

11 A303 [7295 2895]

13 SOUTH CHERITON [692 249]

COMBE THROOP BOREHOLE [7260 2350]

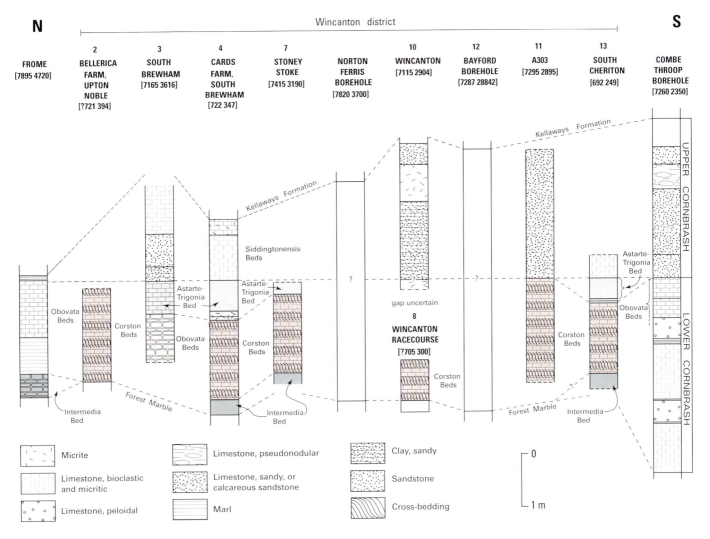

Legend:
- Micrite
- Limestone, bioclastic and micritic
- Limestone, peloidal
- Limestone, pseudonodular
- Limestone, sandy, or calcareous sandstone
- Marl
- Clay, sandy
- Sandstone
- Cross-bedding

0 — 1 m

Figure 22 Comparative sections in the Cornbrash of the district and adjacent areas (see Figure 21 for locations).

UPPER CORNBRASH (MEMBER)

The Upper Cornbrash consists of sparsely sandy, peloidal biomicrite overlying fine-grained, calcareous sandstone and sandy biosparite. The lower part has been called the **Siddingtonensis Beds** (Douglas and Arkell, 1928) after the brachiopod *Microthyridina siddingtonensis*. Nonsequences, including the junction with the Lower Cornbrash, are suggested by erosional contacts and local accumulations of sand and bioturbated shell debris.

Representative sections in the Upper Cornbrash occur near Wanstrow [around 7057 4196 and 7065 4180] (Figure 21, locality 1), where Douglas and Arkell (1928) noted abundant *Aptyxiella lineata*; at Cards Farm (locality 4), South Brewham, where it is composed of 0.3 m of marl and marly ferruginous limestone with *Microthyridina lagenalis*, overlying 0.86 m of hard, purple-centred limestone with burrow trails (possibly Siddingtonensis Beds) (Douglas and Arkell, 1928), and in a bank [7227 3472] (locality 5) north-east of Redlynch, where slightly ooidal,

muddy, coarse-grained, sandy micrite with abundant *M. lagenalis* is exposed. East of Wincanton, exposures [7220 2879 to 7230 2883] (Figure 22, locality 9) showed 2.3 m of shelly micrite in units up to 0.4 m thick, and [at 7295 2895] (locality 11) at least 2.5 m of calcareous sandstones and sandy limestones with *M. siddingtonensis* and *Macrocephalites* cf. *verus*, resting with a sharp break on the undulating top of the Lower Cornbrash (Bristow et al., 1992). Near there, Douglas and Arkell (1928) saw 'concretionary sandy limestones, sandstones and sand' with '*Microthyris siddingtonensis*' and '*Rhynchonelloidea cerealis*' resting on the eroded top of the Lower Cornbrash. At South Cheriton, a quarry [692 249] (locality 15) exposed 0.46 m of hard, purplish limestone with '*M. siddingtonensis*' and '*R. cerealis*', with a basal marly parting, resting on the Astarte–Trigonia Bed (Douglas and Arkell, 1928).

The following newly exposed section [7165 3616] (Figure 21, locality 3) is representative of the Cornbrash in the South Brewham area. The Corston Beds, present at Cards Farm to the north-north-west, are absent:

	Thickness m

UPPER CORNBRASH

Limestone, buff to ochre weathering, locally with pitted appearance, pale brownish grey, fine-grained, sparsely sandy biomicrite, weakly and irregularly bedded: sparsely shelly with some large complete bivalves, locally with up to 10 per cent yellow, medium-grained (to 0.5 mm across) peloids; fine-grained (typically 0.2 mm) quartz grains up to 5 per cent; 30–40 per cent ferroan micritic calcite cement (E70354a)   1.00

Sandstone, pale pinkish purple to buff weathering, pale grey, fine-grained, calcareous, and very calcareous, including sandy biomicrite (E70354b) in upper 0.24 m; wavy bedded: some lignite fragments up to 50 mm, locally with abundant shell debris and burrows; angular quartz grains 0.1–0.3 mm size (fine sand), sharp base   0.65

Limestone, sandy biomicrite, hard and well-cemented, massive, forming conspicuous ledge, some broken shell fragments and whole, sparry-filled, shells; some 30 per cent fine-grained (c. 0.5 mm), angular quartz sand; locally bioturbated and peloidal (E70354c)   0.22

Mudstone, dark brown, sandy, irregular, locally shelly   0.05

LOWER CORNBRASH

Astarte–Trigonia Bed

Biosparite, greyish to pale brown weathering, crowded with bivalves, particularly in the uppermost 0.2 m, including *Meleagrinella echinata*, *Neocrassina hilpertonensis*, *Pholadomya deltoidea*, *Pleuromya subelongata?*, *P. uniformis*, *Radulopecten vagans*, fragments of *Pygurus michelini* and *Kallirhynchia*; shell fragments at least 50 per cent of rock, non-ferroan calcite; locally peloidal, medium or coarse grained (E70354d)   0.65

Obovata Beds (probable equivalent)

Limestone, pseudonodular, pale grey weathering, with mostly fragmented shells; extremely bioturbated with large and indistinct horizontal and vertical burrows up to 10 mm, rubbly textured, common oysters, burrowing bivalves, rhynchonellids; irregular and wispy seams and beds of marl, becoming muddier towards base   0.90

In addition, loose *Cererithyris intermedia* suggest that the Intermedia Bed is close to the floor of the pit.

SIX

Middle (part) and Upper Jurassic

The youngest Middle Jurassic and Upper Jurassic strata crop out in the west and south of the district. They comprise a sequence of clays alternating with less dominant limestones (Tables 8, 9, 10). The ammonite-based standard zones and subzones are treated as chronostrati-graphical units, and the nominal taxa are recorded in roman font.

A progressive marine transgression in Late Bathonian and Early Callovian times established a warm, shallow sea tens of metres deep, surrounded by islands and larger

Table 8 Summary of the stratigraphy of the Kellaways and Oxford Clay formations.

Chronostratigraphical units				Lithostratigraphical units			Lithology
Stage	Substage	Zone	Subzone				
UPPER JURASSIC	OXFORDIAN	Lower Oxfordian	Cordatum*	Cordatum*	Hazelbury Bryan Formation (pars)		Orange-brown, bioturbated muddy sand and sandy mudstone
				Costicardia?	STEWARTBY AND WEYMOUTH MEMBERS (UNDIVIDED)	Red Nodule Beds	Bluish grey, calcareous, shelly mudstone with small (10 cm) cementstone nodules
				Bukowskii?			
		Mariae*	Praecordatum*				
			Scarburgense				
MIDDLE JURASSIC	CALLOVIAN	Upper Callovian	Lamberti	Lamberti	OXFORD CLAY FORMATION		
				Henrici			
		Athleta*	Spinosum?				
			Proniae?		PETERBOROUGH MEMBER	Comptoni Bed	Brown, silty, bituminous shales with some beds of grey mudstone; five beds of large cementstone nodules
			Phaeinum*				
		Middle Callovian	Coronatum*	Grossouvrei*			
				Obductum*			
		Jason*	Jason*				
			Medea*			Mohuns Park Member	Grey silty mudstone
		Lower Callovian	Calloviense*	Enodatum*			
				Calloviense*	Kellaways Formation		Clayey sandstone, silty and sandy mudstone
			Koenigi*	Galilaeii			
				Curtilobus*			
				Gowerianas*			
			Herveyi	Kamptus			
				Terebratus	Cornbrash (Upper)		Sandy biomicrite
				Keppleri			

* Zone/subzone proved in the Wincanton district
? Subzone doubtfully proved in the Wincanton district

landmasses that supplied abundant organic matter. The position of the shorelines is unclear. The significant sand content of the Kellaways Formation implies deposition not too far from the shoreline. During very low sea-level stands, carbonate-cemented, muddy sandstones were deposited (MacQuaker, 1994). In contrast, the relatively homogeneous mudstone lithology of the Oxford Clay, suggests that shorelines were distant from the present outcrop (Hudson and Martill, 1994). Two main facies dominate the Oxford Clay — organic-rich, fissile, very fossiliferous mudstone (Peterborough Member) and calcareous, less organic-rich, blocky, less fossiliferous mudstone (Mohuns Park, Stewartby and Weymouth members). These are thought to reflect episodic deposition controlled by sea-level fluctuations. A recent model (MacQuaker, 1994) proposes that the organic- and silt-rich mudstones formed during periods of relatively low sea-level stand, while more clay-rich, organic-poor mudstones formed during periods of high sea-level stand. In this case, the transition from the Peterborough to Stewartby and Weymouth members may reflect a general deepening of the sea. Sandy mudstones at the top of the Mohuns Park Member may have been deposited during very low, sea-level stands. However, in the latter case, the local extent of this facies suggests that water shallowing may have been tectonically controlled. The model also predicts that during very high sea-level stands, when sedimentation rates were very low, carbonate cements will precipitate in the mudstones, and concretionary carbonate facies occur.

KELLAWAYS FORMATION

The term Kellaways Stone was introduced by William Smith at the beginning of the 19th century for the calcareous sandstones that occurred between the Cornbrash and Oxford Clay near Kellaways in Wiltshire. The name was modified to Kellaways Beds by Woodward (1895). More recently (for example Page, 1989), the formal name Kellaways Formation has been adopted. During the original survey, the formation (then Kellaways Beds) was mapped as part of the Oxford Clay. In the type area, the formation comprises two members, the Kellaways Clay, overlain by the Kellaways Sand. These two members have only been recognised locally in the Wincanton district.

The Kellaways Formation has a faulted outcrop, extending from near Cheriton [c.695 255] in the south to Wanstrow [715 420] in the north. Thickness at outcrop ranges from 15 to 25 m. The Norton Ferris Borehole [7820 3700] proved 22 m. The Gillingham Borehole [7959 3700] proved 21.5 m, but the formation is relatively thin compared to nearby areas in the northern part of the Shaftesbury district (Bristow et al., 1995, figs. 33 and 35); this may be partly due to inadequate data from the Gillingham Borehole for accurately determining the formation boundaries.

The formation is dominated by medium-grey sandy mudstone that weathers to yellow or orange-brown, ferruginous clayey sand and sandy clay. Exposure is generally poor, but there is sufficient evidence to suggest that the sequence is similar to, but thinner than, that in the adjoining Shaftesbury district. The Coombe Throop Borehole [7260 2350] on the northern edge of the Shaftesbury district provides a good representative section of the formation. There, the formation consists of about 2.5 m of sandstone or muddy sandstone, overlain by about 6.5 m of silty mudstone, and about 28 m of sandy mudstones and muddy sandstones with some small cementstone nodules; in the lower part, the **Henstridge Bed** (Barton, 1990), forms a regional marker bed. It is a 0.3 m-thick, laminated, probably storm-deposited, calcareous sandstone, and has been seen in exposures [7198 3052] near Charlton Musgrove and [7254 3627] South Brewham.

Although not shown on the map, the Kellaways Clay and Kellaways Sand members of the Wiltshire type area have been recognised in this district in a cutting at Devenish Hill [726 286] south of Wincanton, and in a section [7175 3050 to 7238 3055] near Charlton Musgrove (Bristow et al., 1993). The Kellaways Clay comprises silty clays and silts with scattered, locally septarian, concretions varying from 0.05 to 1.2 m in diameter. The overlying Kellaways Sand comprises clayey sand and soft sandstone with sandy clay, typically with abundant *Gryphaea (Bilobissa) dilobotes*. The sand member was recognised in stream exposures [7272 449] east of Redlynch. The top of the Kellaways Formation is marked by a rapid change from sandy clay or mudstone to uniform silty mudstone of the Oxford Clay.

The formation ranges from the Herveyi Zone to the Calloviense Zone of the Lower Callovian (Callomon, 1964; Page, 1989) (Table 8); a more refined scheme of ammonite faunal horizons is also available for this interval (Callomon et al., 1989; Page, 1995). The following zones and subzones have been recognised in the district: Koenigi Zone, Gowerianus Subzone (*Cadoceras* aff. *tolype*, *Chamoussetia buckmani*, *Kepplerites (Gowericeras) gowerianus* and *Proplanulites koenigi* from Devenish Hill [726 286], *Macrocephalites* ex gr. *lophopleurus* from [7190 3052] near Charlton Musgrove); also near Charlton Musgrove, Koenigi Zone, Curtilobus Subzone (*Proplanulites* cf. *ferruginosus* from [7223 3054]); Calloviense Zone and Subzone (*Sigaloceras* ex gr. *calloviense* from [7232 3054], *Proplanulites* ex gr. *petrosus* from [7238 3055]) (Bristow et al., 1993).

OXFORD CLAY FORMATION

Over much of England, the Oxford Clay is divided into three members. In ascending sequence, these are the Peterborough Member (formerly the Lower Oxford Clay), Stewartby Member (formerly the Middle Oxford Clay) and the Weymouth Member (formerly the Upper Oxford Clay). The Lamberti Limestone, the top of which defines the base of the Weymouth Member, has not been recognised in this district; the upper two members are therefore shown as a single unit on the map. Locally in the Wincanton and Shaftesbury districts, an additional unit, the Mohuns Park Member has been mapped at the base of the Oxford Clay. The outcrop of the Oxford Clay forms a wide north–south-trending vale across the west side of the district. The formation ranges in thickness from an exceptionally thin 44 m to the south of South Brewham (Freshney, 1995a) to 122 m in the Norton Ferris Borehole; it is up to 135 m in the adjoining Shaftesbury district (Bristow et al., 1995). An anomalous 161 m was recorded in the Gillingham Borehole (Whitaker and Edwards, 1926), but this may be erroneous (see Kellaways Formation above).

The Mohuns Park Member consists of medium grey, weakly calcareous, silty, slightly pyritous, shelly mudstone (Table 8). The Peterborough Member comprises organic-rich ('bituminous') mud-stones. The upper and greater part of the formation (Stewartby and Weymouth members) consists of medium grey, calcareous mudstone. Augering to depths of up to 3 m to unweathered clay is required before the various members can be differentiated.

Mohuns Park Member

The Mohuns Park Member has a narrow, partially faulted, discontinuous outcrop. It varies in thickness from 0 to 15 m, but it is usually between 3 to 10 m thick.

The base of the member is differentiated from the Kellaways Formation by a lack of sand. Typically it consists of grey, silty, shelly mudstone with *Kosmoceras?*, *Procerithium*, *Dicroloma* and small bivalves, including *Corbulomima* and *Bositra buchii*. Such a lithology was seen in an exposure [7330 3679] east of North Brewham. At the top of the member north of this locality, a sandy mudstone, up to 3 m thick, is similar to the Kellaways Formation, [such as between 7338 3683 and 7361 3690]. The mudstones weather to a greyish brown, with much jarositic and ferruginous material. Calcareous concretions occur locally.

The base of the member falls in the Calloviense Zone, Enodatum Subzone. The subzonal index *Sigaloceras enodatum* was recorded at Devenish Hill [726 286] (Bristow et al., 1993). The top is diachronous, ranging from the Jason Zone, Medea Subzone to, maybe, the Coronatum Zone, as in the Shaftesbury district (Bristow et al., 1995). Records of *Kosmoceras jason* [7294 3452 and 7280 3399], south of South Brewham, indicate the Jason Zone and Subzone.

Peterborough Member

The outcrop of the Peterborough Member runs parallel to that of the Mohuns Park Member. The boundary between the two members is gradational over several metres. Across the district, the member tends to form a convex scarp feature; this may be caused partly by several beds of large cementstone nodules (known from the Shaftesbury district (Bristow et al., 1995)) and partly to the slightly firmer nature of the organic-rich 'bituminous' mudstones. The member ranges from 5 to 18 m in thickness.

The Peterborough Member consists of brown fissile mudstones, with some less fissile, medium to dark grey and brownish mudstones. Near the surface, the mudstones weather to pale grey, orange-mottled clays, commonly with fine-grained granular gypsum. The zone of weathering is usually 1.5 to 2 m thick, and the transition to hard, unweathered mudstone commonly takes place over 0.5 to 1 m. The member is the most fossiliferous part of the Oxford Clay; shell beds rich in bivalves, notably nuculaceans or *Meleagrinella*, are relatively common, and are thought to represent non-sequences or to relate to storm events (Hudson and Martill, 1991).

The base of the member is diachronous. In the district, the Jason Zone, Medea Subzone, is the oldest proved. This is indicated by the ammonites *Kosmoceras medea?* and *Pseudocadoceras* from a trench [7250 3055 to 7255 3055] near Charlton Musgrove (Bristow et al., 1993). The Coronatum Zone, Obductum Subzone, is proved by *Kosmoceras obductum* and *Erymnoceras* in a stream [7397 3362], south-east of South Brewham. *Binatisphinctes comptoni* and *Kosmoceras* (including *K. ?ex gr. grossouvrei*) was found in fissile, greyish brown mudstone, associated with very shelly nuculacean-rich mudstone with *Erymnoceras*, in a trench section [7304 3050] near Charlton Musgrove. This fauna indicates the Coronatum Zone, Grossouvrei Subzone, close below the Comptoni Bed; the latter is a nuculacean-rich mudstone which forms a widespread marker horizon across southern and eastern England. *B. comptoni* and looped-ribbed *Kosmoceras* from similar mudstones in a river exposure [7371 3698] south-east of Border indicate the Athleta Zone, Phaeinum Subzone, close above the Comptoni Bed. The Phaeinum Subzone was also proved by *Kosmoceras*, including *K. ex gr. phaeinum* and *K. (Spinikosmokeras)*, in an exposure [7369 3697] east of North Brewham.

Stewartby and Weymouth members

The outcrop of these members runs north to south across the district and is displaced by the Warminster and Witham Friary faults in the north and by the Mere Fault farther south. They also occur in a faulted outcrop south-west of Wincanton [702 277], where the junction with the Hazelbury Bryan Formation was seen (Bristow et al., 1993).

The combined Stewartby and Weymouth members range in thickness from 70 m in the north of the district to 90 m in the Norton Ferris Borehole. A thickness of 114 m in the Gillingham Borehole may represent local thickening into the Mere Basin, but there is some doubt about the interpretation of the borehole sequence (p.45). These thicknesses compare with 85 m on the northern margin of the Shaftesbury district (Bristow et al., 1995). The 30 m recorded in the area south-east of South Brewham (Freshney, 1995a) is exceptional. Freshney (1995a) concluded that strata equivalent to the Stewartby Member were no more than 15 m thick in the Lawrence Hill area [702 277].

The members comprise calcareous, medium grey, variably silty, shelly mudstones, with some very fine-grained sand. In the weathered zone, the beds are yellow or greyish yellow, commonly with gypsum crystals, up to 10 mm long. Some of the commonest fossils are thick-shelled gryphaeid oysters, particularly *Gryphaea dilatata* in the upper part. Cementstone nodules, up to 0.3 m across, occur sparsely at some levels. Near the top, small, red, ferruginous nodules, possibly from the Red Nodule Beds, were seen in a trench [7422 3065 to 7480 3050] near Charlton Musgrove.

The members range from the Athleta Zone, Proniae Subzone, to the Cordatum Zone, Costicardia Subzone

(Table 8). The following zones and subzones have been recognised in the district: Athleta Zone, Proniae or Spinosum Subzone (*Alligaticeras* aff. *rotifer* from a trench [7388 3051] east of Charlton Musgrove): Mariae Zone, Praecordatum Subzone (common *Cardioceras (Scarburgiceras) praecordatum* from a section [701 277] at Lawrence Hill (Bristow et al., 1993)): Mariae Zone, Praecordatum Subzone or Cordatum Zone, Bukowskii Subzone (*Cardioceras (Scarburgiceras) praecordatum* trans. *C. (S.) bukowskii* from a trench [7504 3051] near Charlton Musgrove): Cordatum Zone, ?Costicardia Subzone (*Cardioceras (Subvertebriceras)* sp., *Goliathiceras (Pachycardioceras) anacanthum* and *Peltoceras (Peltomorphites)* aff. *eugenii* from a section [701 277] at Lawrence Hill (Bristow et al., 1993). The Athleta Zone was also proved by palynomorphs and calcareous microfaunas recovered from clays beneath river terrace deposits of the River Cale [7175 2656] (Wilkinson, 1993b, unpublished report*).

CORALLIAN GROUP

The Corallian Group represents an episode of shallow-marine, mixed carbonate and siliciclastic sedimentation which interupted a long period of deeper-water, argillaceous, shelf sedimentation that started with the Kellaways Formation and ended with the Kimmeridge Clay. The group comprises three broad divisions. The lower part consists of sandy clay and fine-grained sand of the Hazelbury Bryan Formation. There is a median unit, Stour Formation and Clavellata Beds, of commonly ooidal, calcareous clays, and ooidal, pisoidal and bioclastic limestones. The upper unit is a mixture of sandy and argillaceous beds of the Sandsfoot Formation, Ringstead Waxy Clay and 'Ampthill Clay'. In the Wincanton district and in the Shaftesbury district to the south, the Ringstead Waxy Clay and Ampthill Clay are not easily distinguished by augering from the Kimmeridge Clay and have been mapped with the last formation. The stratigraphical sequence of the district (Table 9) can be traced throughout Dorset.

The Corallian Group crops out mainly to the south of the Mere Fault. North of the fault, only the Hazelbury Bryan Formation and, very locally, the basal part of the Stour Formation, is preserved beneath unconformable Cretaceous strata, except near the northern margin of the district, just north of Horningsham [around 815 428], where almost the whole succession crops out.

The formation thickens south-eastwards from about 30 to 80 m (Figure 23). The complete succession was penetrated in the Wyke Brewery Borehole [7959 2661], where 74.4 m was proved.

In the mid and late Oxfordian, ammonite provincialism became acute and different standard zonations have been developed for each province. The one traditionally applied to the Corallian Group of Dorset is based mainly on perisphinctids, but a more boreal zonation based exclusively on cardioceratids, used elsewhere in England, can also be applied in part (Table 9). The group ranges from the Lower (Cordatum Zone, Cordatum Subzone) to Upper Oxfordian (Pseudocordata Zone).

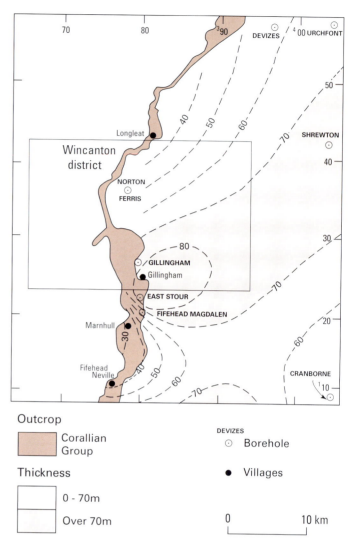

Figure 23 Isopachyte map for the Corallian Group (contours at 10 m intervals).

Hazelbury Bryan Formation

The formational name was introduced by Bristow (1989a) as a replacement for the inappropriately named 'Lower Calcareous Grit'. The type area is the village of Hazelbury Bryan in the Shaftesbury district.

The formation crops out across the whole district, and is offset by the Mere Fault. There is a small outlier [702 277] south-west of Wincanton, and a very small inlier in the core [7860 3123] of the anticline south-east of Zeals (Bristow et al., 1992).

In the north of the district, the formation varies rapidly from 15 to 40 m; it is about 20 m thick north of Horningsham. An incomplete 25 m (top missing beneath the unconformable Gault) was proved in the Norton Ferris Borehole, and 28.4 m was proved in the Gillingham Borehole. The formation is about 35 m thick on the southern margin of the district (Henderson et al., 1994), and to the south, 52.5 m was proved in the Fifehead Magdalen Borehole [7985 2100] (Bristow, 1990a).

Table 9 Chronostratigraphical subdivision of the Oxfordian Stage as applied to the Corallian Group.

Predominantly Perisphinctid zonation		Substage	Cardioceratid zonation		Lithostratigraphy of the Corallian Group of the Wincanton district		
Zone	Subzone		Subzone	Zone			
Pseudocordata	Evoluta	Upper Oxfordian		Rosenkrantzi	'Ampthill Clay' Ringstead Waxy Clay		
	Pseudocordata						
	Pseudoyo			Regulare	Sandsfoot Grit		Sandsfoot Formation
	Caledonica						
Cautisnigrae	Variocostatus						
	Cautisnigrae		Serratum	Serratum			
Pumilus	Nunningtonense		Koldeweyense		Sandsfoot Clay		
			Glosense	Glosense	Clavellata Beds	Ecclife Mb	
			Ilovaiskii				
	Parandieri	Middle Oxfordian	Blakei	Tenuiserratum			
Plicatilis	Antecedens		Tenuiserratum		Newton Clay Sturminster Pisolite		Stour Formation
			Maltonense	Densiplicatum			
	Vertebrale		Vertebrale		Cucklington Oolite Woodrow Clay		
Cordatum	Cordatum	Lower Oxfordian	Cordatum	Cordatum	Hazelbury Bryan Formation		
	Costicardia		Costicardia		Oxford Clay (Weymouth Member)		
	Bukowskii		Bukowskii				
Mariae	Praecordatum		Praecordatum	Mariae			
	Scarburgense		Scarburgense				

(Right-hand column spanning the lithostratigraphy: Corallian Group)

The zonations are based on ammonite faunas — one exclusively cardioceratid (for the boreal and subboreal areas) and one predominantly perisphinctid (for parts of the subboreal area) (Sykes and Callomon, 1979, emend. Birkelund and Callomon, 1985; Wright, 1980).

The formation consists mainly of clay and sandy clay, with clayey sand, sand, and thin sandy limestone mostly at the top of coarsening-upward sequences (Figure 24). In addition, ooidal clay occurs near the top of the formation [7577 2625 and 7613 2564], and is locally associated with shelly limestone, and limonite nodules [7623 2564 and 7615 2602]; calcareous, shelly doggers [7636 2590] occur near Cucklington. Smectite is the dominant clay mineral, but kaolinite is also common, mainly in the sandier parts of the sequence (Bristow et al., 1995).

The sands range from poor to well sorted and are very fine to medium grained (Bristow et al., 1995). Some are calcareously cemented and ooidal. Generally, their bases are marked by spring lines [for example around 7656 2380]. South of the Mere Fault, there are four sand units (A, B, C and D; Figure 24), but all four sands are not developed everywhere. In the north, near Horningsham, thin impersistent beds of fine-grained sand occur [for example 8098 4236 to 8108 4213 to 8122 4222], but these cannot be correlated with sands A to D.

South of the district, the base of the formation is defined by the base of Sand A, but in this district, Sand A is mostly missing, and the base is taken at the upward change from smooth clays of the Oxford Clay to very sandy clays of the Hazelbury Bryan Formation; the base is commonly marked by a weak spring. Sand A occurs at Lawrence Hill [701 277], where Professor J H Callomon recorded silty clays and silts passing up into yellow doggery sands (Bris-

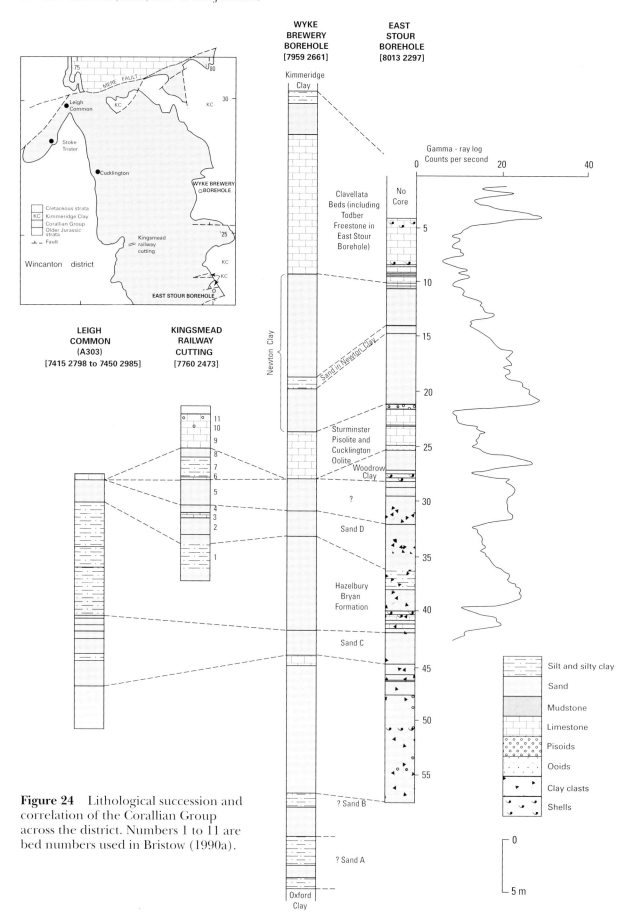

Figure 24 Lithological succession and correlation of the Corallian Group across the district. Numbers 1 to 11 are bed numbers used in Bristow (1990a).

tow et al., 1993). These sands were augered nearby [7020 2271] and were seen in pits [7022 2775] north of the A303 by Woodward (1895), who noted 4.5 to 6 m of sands with concretionary masses of stone, overlying shelly sandy clay with '*Ammonites cordatus*'.

Sand C was well exposed in cuttings [7415 2978 to 7450 2985] for the A303 at Leigh Common (see below). A lenticular bed, up to 4 m thick, of fine-grained sand that crops out over some 400 m [7657 2351 to 7660 2385] south of Buckhorn Weston may be Sand C.

Sand D is more persistent than Sand C. It extends all the way round the outcrop at Stoke Trister where it occurs at the very top of the formation. Cuttings [7415 2978 to 7450 2985] at Leigh Common on the A303 (upper part of section slightly modified after Bristow et al., 1992) provide a representative section in the top of the formation:

	Thickness m
Cucklington Oolite	unexposed
HAZELBURY BRYAN FORMATION (uppermost 6 m unexposed, but proved by augering)	
Sand (Sand D), fine-grained	c.2.00
Sandy clay	c.4.00
Clay, orange-brown, sandy, with limonitic ironstone nodules	?2.00
Clay, bluish grey; sandy, becoming a silty sand towards the base; shelly with bivalves in solid preservation at base including *Isognomon* sp., *Modiolus bipartitus*, *Nanogyra nana*, *Pleuromya alduini* and *Thracia depressa*; fragments of *Cardioceras* including *C.(C.)* ex gr. *ashtonense-persecans*	4.50
Sand, ferruginous with limonitic ironstone nodules (top of Sand C)	0.20
Clay, silty with oysters	0.39
Silt and fine-grained sand, orange-brown with serpulids, *Modiolus bipartitus* and fragment of small belemnite (*Hibolithes?*)	0.90
Clay, grey	0.50
Sand, ferruginous, weakly cemented at top with ammonite fragment (*Cardioceras?*); weakly cemented sandstone 0.15 m-thick, 0.5 m from the top, with *Lopha genuflecta*, *Modiolus bipartitus*, *Cardioceras* spp. including *C.(C.)* ex gr. *ashtonense-persecans*, *C. (C.* trans. *Vertebriceras*) and *C. (Plasmatoceras)* ex gr. *plasticum*	0.77
Sand, silty, becoming less sandy downwards; broken shell fragments (possibly the base of Sand C)	1.80
Clay, medium grey, silty, some fine-grained sand	4.00+

Southwards from Clapton Farm [751 293] to Cucklington, in an area much obscured by landslip, a lenticular clay, up to 3 m thick, and with quite a wide outcrop on the interfluve, occurs between Sand D and the overlying Stour Formation (Bed 5 in the section below). Between Cucklington and Buckhorn Weston, Sand D is seen as clayey fine-grained sand in landslip scars [7554 2782 and 7569 2684] and burrows [7580 2661 and 7659 2635]. It crops out in the sides of the Kingsmead railway cutting (Bed 4) (Bristow, 1990a) (Figure 24):

	Thickness m
TOPSOIL	0.76
STOUR FORMATION (Figure 23, beds 7–11)	
HAZELBURY BRYAN FORMATION	
5. Clay, hard, bluish grey, silty	2.44
4. Sand (Sand D), buff-grey, orange top, fine-grained, silty	0.76
3. Limestone, sandy, dark [grey?], compact, bedded, *Pleuromya alduini* (approximate level of roof of tunnel)	0.46
2. Sand, buff, fine-grained, silty	1.52
1. Sand, bluish grey, fine-grained, or silt, alternating with buff silt in top 0.9 m (base Sand D), passing downwards into clayey silt with dark bituminous clay layers near base. Fossils include *Cardioceras (Plasmatoceras)* aff. *tenuicostatum*	4.27

Bed 1 is part of a coarsening-upward sequence that terminates in sand of Bed 4 (Figure 24). Springs from the base of the sand in Bed 1 caused serious problems during the tunnel construction (Fraser, 1863). Blake and Hudleston (1877) recorded 'immense spheroidal doggers of calcareous grit' in Bed 1, some of which can still be seen in the sides of the cutting.

The macrofauna of the formation is dominated by bivalves, including *Chlamys*, *Gervillella*, *Isognomon*, *Lopha*, *Modiolus*, *Myophorella*, *Pinna* and *Thracia*; gastropods, such as *Dicroloma* and *Procerithium*, are also found. *Cardioceras cordatum?* from Sand B in the Hazelbury Bryan Borehole, south of the district, indicates the Cordatum Subzone (Bristow et al., 1995). Ammonites from Sand C in the Leigh Common section on the A303 (see above) also indicate the Cordatum Subzone. *Cardioceras (Plasmatoceras)* ex gr. *tenuicostatum*, indicative of the Vertebrale Subzone, was found in Sand D in the Kingsmead railway cutting [7760 2473] (Bristow, 1990a). The Cordatum–Vertebrale subzonal boundary therefore occurs in, or at the top of, Sand C. Faunas of the Vertebrale Subzone also occur in the Cucklington Oolite indicating that the upper part of the Hazelbury Bryan Formation belongs exclusively to that Subzone.

Stour Formation

The term Stour Formation as used by Bristow, (1989b) is a modification of Wright's (1980) formation of the same name (see Bristow et al., 1995, p.67). The formation as now recognised comprises six members (Table 9), not all of which occur in any one area. They form a sequence of alternating calcareous clays and limestones. The formation crops out mainly south of the Mere Fault, but there is a small outcrop in the north near Horningsham and a very small inlier [762 335] west-south-west of Stourton.

WOODROW CLAY (MEMBER)

The name Woodrow Clay was introduced by Wright (1980) for 2 to 5 m of grey clay at the base of the Stour Formation. At the type locality [759 108] in the Shaftesbury district, the member consists of up to 5 m of grey, generally sand-free, calcareous, locally ooidal clay. The East Stour Borehole [8013 2297], just south of the district, proved some 2.7 m of Woodrow Clay comprising very shelly, slightly sandy clay with scattered ooids, and with a shelly micrite towards the base. The micrite forms a very distinctive peak on the gamma-ray log because of its high mudstone clast content (Figure 24). A similar sequence was recorded in the Bowden Farm Borehole [7741 2364] just south of the district (Henderson et al., 1994), and in the Kingsmead railway cuttings (Figure 24). Both the Woodrow Clay and the higher Newton Clay have fewer ooids and more fine-grained sand than in the type area to the south. In the absence of good exposure, such as in Kingsmead railway cutting, it is difficult to separate the Woodrow Clay from the Hazelbury Bryan Formation. The clay mineralogy of the member varies considerably (Bristow et al., 1995).

Ooidal grey clay was noted between Cucklington and Buckhorn Weston, where it is 1 to 2 m thick and locally passes into sandy clay [763 270; 760 265, around 760 252 and 762 268], and near Cucklington [7553 2901], but these occurrences are too thin (and are possibly lenticular) to map separately. However, close to the Mere Fault, boreholes alongside the A303 proved more than 6 m of ooidal marl beneath Cucklington Oolite, but this ooidal marl was not noted at outcrop nearby. The following borehole is representative [7545 2995]:

	Thickness m
Limestone rubble	1.25
CUCKLINGTON OOLITE	
Limestone, ooidal and shelly, sandy, micritic matrix, scattered mud streaks	3.13
Marl, grey	0.15
Limestone, sandy and muddy, micritic matrix; scattered ooids and shell fragments	0.65
WOODROW CLAY	
Marl, greyish brown	0.45
No core	1.37
Marl, greyish brown, with some well-cemented levels	2.33
Marl, olive-black with abundant ooids	1.20
No core	0.70
Marl, olive-black with abundant ooids	0.77

The macrofauna consists mainly of bivalves (for example, *Chlamys*, *Corbulomima*, *Pinna* and *Pseudolimea*).

CUCKLINGTON OOLITE (MEMBER)

The Cucklington Oolite (Wright, 1980; 1981) takes its name from the small village [756 275] in the south of the district. It crops out locally north of Horningsham; the main outcrop extends from the Mere Fault to the southern margin of the district. There are small outcrops on the hill at Stoke Trister and in the centre of the anticline south-east of Zeals. North of Horningsham, up to 6 m of shelly ooidal sand and sandstone, sandy oolite, crumbly oolite, and soft, marly, pisoidal oolite occur. South of the Mere Fault, the member consists of flaggy, ooidal, bioclastic limestone that varies in thickness mostly in the range 2 to 5 m. An apparent thickness of 15 m on the hill

at Stoke Trister (Freshney, 1995a) is probably a result of cambering.

The Cucklington Oolite at the type locality [7575 2740] was described by Blake and Hudleston (1877), and Bartlett and Scanes (1917, pl. 23a) as 0.7 m of rubbly, oolite and oolitic marl with scattered pisoliths and 'Ammonites plicatilis, Turbo sp., Lima rigida, E. scutatus, sponge', on 0.70 m of black and brown clay with common 'Exogyra nana', on 1.83 m of blue-hearted, gritty, solid oolite.

An extant representative section [7325 2838] was seen on the hill at Stoke Trister:

	Thickness m
CUCKLINGTON OOLITE	
Oolite, creamy yellow, very rubbly and disturbed in an open framework of slabs which dip about 40° down slope. Two lenses of clay dip across the unit at about 20° SE. Other shear planes dip north in a complementary direction. Some inter-slab cavities near the top contain pisoids.	2.0–2.5
Clay, very ferruginous with clasts of limestone	0.3
Clay, grey, silty	0.2
Sandstone, ooidal	over 0.2

Fossils, usually fragmented, are common and include *Nucleolites scutatus*, bivalves, gastropods and rare ammonites. Wright (1982) recorded *Perisphinctes (Arisphinctes) kingstonensis*, indicative of the Vertebrale Subzone, from near Stoke Trister [740 286]. *Aspidoceras [Euaspidoceras]* cf. *catena*, also indicative of the Vertebrale Subzone, was found near Longleat, north of the district (Bristow, 1994).

STURMINSTER PISOLITE (MEMBER)

The name Sturminster Pisolite was formally named by Wright (1980) as a replacement name for part of the 'Marl and Pisolite Series' of Blake and Hudleston (1877) and Pisolite Facies of Gutmann (1970). The type locality is the road cutting [7830 1348] at Sturminster Newton in the Shaftesbury district.

The Sturminster Pisolite has a relatively wide outcrop around Cucklington, but it becomes very narrow along the steep valley sides to the south. In addition, there is a small outlier south-west of Stoke Trister [733 282], small faulted outcrops close to the Mere Fault [around 7480 2987], and a very small, partially fault-bounded, inlier [7860 3123] south-east of Zeals. In the Corallian Group outcrop north of Horningsham, there appears to be no discrete, mappable, unit of pisolite, although pisoids have been noted at several localities [8046 4237, 8069 4249 and 8128 4256]. Generally, the outcrop is marked by common pisoids and pisolite in brash.

In places, the Sturminster Pisolite is grain supported with pisoids up to 8 mm across; elsewhere, the pisoids are matrix supported, commonly in a clayey ooidal micrite. Where grain supported, the member has a maximum-thickness in the district of 0.6 m, but where it is matrix supported, the bed may be up to 3 m thick. At the type section, the member is 1.2 m thick; but locally in that district, it forms a 2 m-thick limestone, consisting mainly of oolite, but with pisolite at the top and bottom (Bristow et al., 1995). In the Wincanton district, near Leigh Common, a borehole [7497 2987] proved 2.25 m of ooidal, muddy limestone with scattered pisoids, on 0.1 m of calcareous mudstone with scattered ooids and pisoids.

At Cucklington, there is a 0.3 m-thick unit of coarse-grained, shelly, flaggy oolite above the pisolite. Oolite above pisolite also occurs along the Bainly Brook, and at the top of the railway cutting [7858 2495], Bugley, where a 1.3 m-thick, fine-grained, cross-bedded, flaggy oolite is overlain by a 20 mm-thick, cross-bedded, fine-grained sandstone of the Newton Clay (Bristow, 1990a).

Although some pisoids are spherical, most are flattened, with their thickness, in extreme cases, being only one quarter of their diameter (Bartlett and Scanes, 1917, pl. 23b; Bristow et al., 1995, pl. 7C). In hand specimen, broken pisoids show a clear concentric arrangement (see Edmunds, 1938, fig.11b). Thin sections (Bristow et al., 1995, pl. 7C) show pisoids set in a matrix of poorly sorted, patchily micritised ooids in a fine-grained ferroan calcite spar cement. The pisoids are irregularly laminated, commonly with a bioclast as their nucleus. Under high magnification, the pisoids show a vermiform internal structure. Wright (1981, p.24) noted that the larger (7.5 to 10 mm), ovoid pisoids are oncoliths of algal origin, whereas the smaller (3 to 5 mm) spherical bodies developed round ooids are true pisoliths.

Fossils are rare, but commonly include the burrowing echinoid *Nucleolites scutatus*. Gutmann (1970) recorded 21 species of gastropods, bivalves and echinoids from the pisolite near Todber in the Shaftesbury district. Ammonites are more common in the Sturminster Pisolite than in the other Corallian members. In general, they are poorly preserved perisphinctids, but include '*Perisphinctes (Dichotomoceras) sp. cf. antecedens*' from just above the 'famous Cucklington Pisolite' at Cucklington (Mottram, 1957, p.163) which is probably indicative of the Antecedens Subzone.

NEWTON CLAY (MEMBER)

The name Newton Clay (Bristow, 1989b) is a widely developed clay unit above the Sturminster Pisolite; it approximates to Arkell's (1933) 'Littlemore Clay in north Dorset'. The type locality is the road cutting [7825 1347] at Newton in the Shaftesbury district.

The Newton Clay crops out in a limited area in the north of the district, and more widely south of the Mere Fault. There is a small anticlinal inlier [7875 3115] south-east of Zeals.

The Newton Clay is one of the thickest members of the Stour Formation, ranging from 5 to 15 m, and varying in thickness abruptly from place to place. It is about 6 m thick in the north of the district. The Gillingham Borehole [7959 2661] proved 14.6 m. It is 5 m thick along Bainly Bottom [770 272 to 780 272], 10 m thick along the West Brook [782 255], up to 10 m thick at Sandley [around 775 245], about 5 m thick on the east bank of the River Stour [around 7885 2420], and a little farther south [around 7920 2325], it is about 10 m. The Hallett's Farm Borehole [8023 2393] proved about 6 m (Henderson et al., 1994).

North of Horningsham, the member consists mostly of ooidal grey clay, but ooidal marl, pale grey marly clay, smooth grey clay, pale pinkish brown clay and grey sandy clay also occur. South of the Mere Fault, it is dominantly a silty and sandy clay that comprises, essentially, one cycle of deposition, in which the sand content increase upwards, although there are minor oscillations within the cycle (Figure 24). Bedding planes are commonly coated with fine-grained sand, of which there are also thin interbeds. Many other lithologies occur: scattered pisoids occur in the basal 0.10 m of sandy mudstone in the A303 cutting [7506 2990] (Bristow et al., 1992): flaggy, fine-grained sandstone occurs at the base [7765 2718] on the north side of Bainly Bottom and near Bugley [7898 2495]: on the outlier north-east of Cucklington, a thin, muddy, shelly, locally ooidal, limestone occurs about 2 to 3 m above the base, and is overlain by slightly more ooidal clay: marly limestone, in sandy, locally ooidal, clay occurs over a 400 m tract [7604 2920 to 7589 2956] near West Bourton: shelly, ooidal, grey clay is the commonest lithology near Silton [around 775 290]: fine- to medium-grained orange sand occurs locally at the top of the member on the south side of the brook [7683 2678 and 7691 2681]: doggers of fine-grained sandstone, up to 1.5 by 1.0 by 0.2 m, occur near the top at Silton [7696 2883]: slabs of fine-grained sandstone and hard, flaggy oolite occur in brash [7737 2894] near Slait Barn, about 5 m below the top: marly limestones in the uppermost part form a transition to the overlying Clavellata Beds in most areas. Fine-grained sand that is thick enough to map separately [e.g. around 774 245] probably correspond to the mid-cycle sand in the Newton Clay of the East Stour Borehole (Figure 24). Burrowing and bioturbation is common throughout; disseminated lignite occurs locally.

In the Shaftesbury district to the south, the clay mineralogy varies from dominantly smectite in the East Stour and Cannings Court boreholes, to dominantly illite in the Hazelbury Bryan Borehole (Bristow et al., 1995). Sands consist of subrounded quartz grains, micritised ooids and common foraminifera set in a mixed ferroan and non-ferroan calcite cement (Bristow et al., 1995). The Newton Clay shows a higher sand content and sparser ooids in this district than to the south.

Shells, both whole and fragmentary, but zonally undiagnostic, are common at certain levels (Blake and Hudleston, 1877, p.276; Bristow et al., 1995). The member is assigned to the Antecedens Subzone, as ammonites indicative of that subzone occur in both the underlying Sturminster Pisolite and overlying Todber Freestone.

TODBER FREESTONE (MEMBER)

The Todber Freestone (Wright, 1980; 1981) is named from the small village in the northern part of the Shaftesbury district. This, the uppermost member of the Stour Formation, has not been recognised with certainty at outcrop in the Wincanton district. Some 3 m was recorded in the Hallett's Farm Borehole [8023 2393] (Henderson et al., 1994) in the south of the district. Wright (1981; 1985) thought that the Todber Freestone north of East Stour passed into a spicular, shelly micrite. As the micrite is so different from the Todber Freestone, and more closely resembles the Clavellata Beds, it is here included in the latter formation.

Clavellata Beds (Formation)

The term Clavellata Beds was introduced by Bristow (1990a) as a replacement name for the '*Trigonia clavellata* Beds' of previous authors. Locally at the top of the succession is the Eccliffe Member (Bristow, 1990a) named after the hamlet [798 253] south of Gillingham where there are reasonable exposures. The member forms a distinct mappable unit which has been traced from West Bourton to East Stour in the south of the district. It closely resembles the Todber Freestone, with which it has been confused (Wright, 1981, p.25). Blake and Hudleston (1877) commented on the similarity of these beds in the Preston (now Pierston) quarry [7950 2828] to the Osmington Oolite [equivalent to the Todber Freestone] of the Dorset coast.

This widespread formation can be traced across the Shaftesbury and Wincanton districts as far as the Mere Fault. There are small fault-bounded anticlinal inliers south-east [787 310] and east [799 317] of Zeals. North of Horningsham, a small outcrop of shelly, locally pisoidal [8032 4207, 8046 4237, 8069 4249 and 8146 4223], oolite and micrite forms well-featured ground [e.g. around 8017 4216 and 8032 4207].

The formation varies from 3 to 15 m in thickness. In the north, it is about 3 m near Horningsham. South of the Mere Fault, 13.2 m were proved in the Gillingham Borehole, and it is at least 10 m thick near Silton. The Hallett's Farm Borehole proved about 13 m (Henderson et al., 1994) of which about 9 m appears to be the Eccliffe Member. The thickest sequences occur where the Eccliffe Member is well developed.

The basal bed and much of the higher part of the Clavellata Beds south of the Mere Fault consists of tough, coarsely bioclastic, ooidal limestone. However, beds of fine- to medium-grained oolite also occur in units up to 0.3 m thick; in hand specimen, these are indistinguishable from the Todber Freestone. Beds of shelly, patchily ooidal, spicular micrite also occur. Beds of sandy marl are common and, locally, they can make up almost half the succession. The limestones range from biosparites and oobiosparites to pelmicrites and biopelmicrites. The bioclasts are either ferroan calcite fragments with micritised envelopes, or unaltered, non-ferroan calcite fragments. Calcite-filled spheres, possibly after *Rhaxella* sponge spicules, together with foraminifera are common. Very fine-grained quartz sand is also common and may form the nuclei of ooids. A representative section [7631 2918] near West Bourton, in an area where the formation is about 10 m thick, shows:

	Thickness m
Topsoil, stony clay	0.30
Limestone, marly	0.10
Marl, buff	0.15
Limestone, marly, crumbly	0.25
Marl, buff with *Bourguetia*, *Natica*, *Chlamys* sp., *Discomiltha rotundata*, *Lopha (Actinostreon) solitaria*, *Nanogyra nana* and *Nucleolites scutatus*	0.20
Limestone, hard, shelly, sparsely ooidal in two beds, each about 0.2 m thick; with *Bourguetia*, *Natica*,	

	Thickness m
Pseudomelania?, *Discomiltha lirata*, *D. rotundata*, *?Nanogyra nana*, *Pholadomya aequalis*, *Pleuromya uniformis* and *Nucleolites scutatus*	0.40
Limestone, sparsely ooidal and shelly	1.50
?gap	
Limestone, nodular, marly partings, sparsely ooidal with sugary texture, shelly with *Pleurotomaria*, *Discomiltha rotundata* and *Pleuromya uniformis:*	0.10
Marl, shelly	0.05
Limestone as above	0.10

A 10 m section in Clavellata Beds, resting on Newton Clay, at Manor Farm, Silton [779 293] was described by Bartlett and Scanes (1917) and Edmunds (1938). Fossils from the Pope-Bartlett bequest in the BGS collections are listed in Bristow (1990a). Arkell (1929) found the first British specimen of *Mytilus varians* in this pit, where it forms a 'mussel bank' (Mottram, 1957).

In the 'new' quarry [7805 2841], Silton, Wright (1985) claimed that 5 m of 'Todber Freestone' were present; all the beds are now assigned to the Clavellata Beds. A section [7805 2841] measured in 1989 is as follows:

	Thickness m
Topsoil and rubbly stone	0.50
Sparry oolite, irregularly flaggy, fossiliferous with common *Myophorella*, *Pseudomelania* and *Perisphinctes*	1.00
Micrite, hard, bluish grey, shelly	0.40
Sand, very fine-grained, clayey with sandstone nodules up to 0.2 m across	0.30
Micrite, hard, bluish grey, shelly	0.50
Marl, bluish grey, silty and very fine-grained sandy, with common *Natica*, and *Pleurotomaria*, *Pseudomelania*, *Deltoideum delta*, *Nanogyra nana*, *Pholadomya aequalis*, *Pleuromya uniformis*, *?Thracia depressa*; bioturbated and burrowed	0.30
Probable small gap in section	
Section continues at [7807 2844]	
Oolite, rubbly, marly	0.15
Oolite, shelly, nodular	0.10
Silt, slightly sandy	0.10
Limestone, irregular nodular	up to 0.10
Silt, sandy	0.13
Limestone, irregular nodular	up to 0.06
Marl, sandy with *Pleuromya*	0.08
Oosparite, hard, shelly	0.40
Oomicrite, shelly, rubbly weathering	1.80

Eccliffe Member

The Eccliffe Member consists of flaggy, fine-grained oolite, and is up to 9 m thick. West of Thorngrove, a quarry [7913 2588], partially filled, exposes 2.4 m of the Eccliffe Member on 1.15 m of undifferentiated Clavellata Beds (Plate 7); and at Eccliffe, a quarry [7997 2503] shows 1.5 m of flaggy, fine-grained oolite passing down into thicker bedded (up to 0.25 m), fine-grained oolite.

The fauna of the Clavellata Beds is dominated by bivalves, gastropods and echinoids (Bristow et al., 1995; Gutmann, 1970). Corals, including *Stylina*, occur in masses up to 0.15 m diameter; *Thamnastraea* and *Thecosmilia* occur

Plate 7 Junction of the generally flaggy, fine-grained, sparsely bioclastic oolite of the Eccliffe Member of the Clavellata Beds (upper third of photograph) with more massive, bioclastic limestones interbedded with sandy marls of the undifferentiated Clavellata Beds. Hammer length 30 cm. (A15117)

locally [7950 2828] (Arkell, 1933; Blake and Hudleston, 1877). From the 'old' quarry at Whistley Farm [7815 2850], Arkell (*in* Mottram, 1957) identified '*Decipia* aff. *lintonensis, D.* cf. *decipiens, Perisphinctes (Pseudarisphinctes)* cf. *shortlakensis, P.* sp. indet. cf. aff. *decurrens* and *Perisphinctes* sp.'. Wright (1985) recorded many ammonites, including *Perisphinctes* spp. and *Amoeboceras glosense*, from the 'new' quarry at Whistley Farm. Edmunds (1938) recorded '*Perisphinctes plicatilis*' in the quarry [779 293] at Silton. *Decipia* sp. was found in the Clavellata Beds of the Zeals Park inlier [around 800 318]. The ammonites allow a good correlation between the Clavellata Beds of this district and the coast, and indicate the Cautisnigrae Subzone. '*Perisphinctes* cf. *antecedens*' found in brash [787 311] on Clavellata Beds at Wolverton (Mottram, 1957), may suggest that the base of the formation is older in the present district than it is in the Shaftesbury district to the south (Bristow et al., 1995).

Sandsfoot Formation

The name of the formation, as Sandsfoot Series, was introduced by Blake and Hudleston (1877) based on the coastal exposures at Sandsfoot, near Weymouth. The unit was given formational status by Wright (1980).

The formation extends southwards from the Mere Fault to the district margin. In the north, there is a small outcrop north of Horningsham.

South of the Mere Fault, the formation thins southwards from 10 m north and north-west of West Bourton [761 300 to 765 295], to about 8 m at Silton and near Milton on Stour, to about 2 m in the area west of Wyke and to about 1.5 m south and east of the River Stour [around 798 240]. The Gillingham Borehole [7959 2661] proved 3.8 m.

The Sandsfoot Formation is locally divided into the Sandsfoot Clay overlain by the Sandsfoot Grit.

SANDSFOOT CLAY (MEMBER)

The Sandsfoot Clay consists of grey, ooidal, locally sandy, shelly and ferruginous clay. It crops out only in the area south-east of Bourton [around 777 297], where it is up to 3 m thick. A borehole [7731 2979] proved the following succession:

	Thickness m
TOPSOIL	0.20
SANDSFOOT GRIT	
Sand, brown, fine- and medium-grained, clayey, becoming more clayey with depth, medium dense	0.60
SANDSFOOT CLAY	
Clay, firm to stiff, pale grey, mottled orange, fine-grained sandy	1.40
Clay, firm, orange-brown, scattered shell fragments, some ironstone nodules	0.40
Clay, very stiff, dark grey, silty, fissured with abundant shell fragments	0.60

In the Hallett's Farm Borehole [8023 2393], the member consists of about 1.6 m of silty clay (Henderson et al., 1994).

SANDSFOOT GRIT (MEMBER)

The member varies from a fine-grained sand, commonly ferruginous, to a fine-grained sandy clay. It becomes much more clayey as traced from north to south; locally it passes into a very sandy clay. In places such as south of Feltham Farm, interbedded fine-grained sandy and silty clay with shell fragments, up to 1.1 m thick, has been proved in boreholes [7734 2988]. Augering west of Park Hill, Longleat, proved 3 m of brown, ferruginous loam with black limonite ooids [8172 4291].

In many areas, small oysters are common; other bivalves, including *Ctenostreon, Deltoideum delta, Discomiltha* and *Pleuromya*, serpulids and belemnites also occur (Bristow et al., 1992). The Sandsfoot Grit on the Dorset coast has yielded perisphinctids and *Amoeboceras*, indicative of the Pseudoyo and Pseudocordata subzones (Wright, 1986).

Ringstead Waxy Clay and 'Ampthill Clay' (formations)

On the Dorset coast, the Ringstead Waxy Clay (Arkell, 1947), capped by the Ringstead Coral Bed, separates the

Sandsfoot Formation from the Kimmeridge Clay. In this district, and in the Shaftesbury district to the south, a second unit, the 'Ampthill Clay' (Seeley, 1869) is present between the Ringstead Waxy Clay and Kimmeridge Clay. They can be distinguished in extended sections and cored boreholes, but cannot be satisfactorily separated in mapping, and are combined under the name Kimmeridge Clay on Sheet 297 (Wincanton) and Sheet 313 (Shaftesbury).

The thickness of the Ringstead Waxy Clay and 'Ampthill Clay', both of which are of Late Oxfordian age, is estimated to be about 25 m at Gillingham.

The **Ringstead Waxy Clay** is characteristically pale grey, generally smooth-textured, poorly fossiliferous, and with red sideritic nodules. It was proved at the sewage works (Figure 25, locality 1) on the western outskirts of Gillingham.

The '**Ampthill Clay**' consists mainly of dark grey, sparsely to moderately shelly mudstones. There may be several interburrowed horizons, and interbedded paler grey mudstones. The fauna is dominated by bivalves, especially *Deltoideum delta*; ammonites include partially pyritised and iridescent 'perisphinctid' nuclei and macroconch *Ringsteadia* (commonly found only as smooth body-chamber fragments). The unit almost certainly warrants a new name, but, pending more work on the regional stratigraphy, it is provisionally assigned to the 'Ampthill Clay', a formation developed in similar facies at this stratigraphical level in central and eastern England (Gallois and Cox, 1977). In recent accounts of sections in the Wincanton district (e.g. Bristow et al., 1992), the unit was classified as the upper part of the Ringstead Waxy Clay, following the terminology used in the adjoining Shaftesbury district (Bristow et al., 1995).

'Ampthill Clay' was exposed along the Gillingham inner relief road (Figure 25, locality 5), at Colesbrook (locality 6), along the A303 near Zeals (locality 9), as well as at other smaller sites west of Gillingham (localities 2, 3 and 4) (Bristow, 1993; Bristow et al., 1992). It was also probably exposed in the former Gillingham brickpit (locality 7) (Woodward (1895), Pringle *in* White (1923), Arkell (1933) and Edmunds (1938).

The boundary with the Ringstead Waxy Clay at Gillingham sewage works (locality 1) is marked by a distinctive horizon of interburrowing with associated clay ironstone and phosphatic nodules. Near the base of the 'Ampthill Clay', there is usually a shell bed, locally cemented, which has yielded diverse bivalves, including *Camptonectes*, *Chlamys*, *Corbulomima*, abundant *Deltoideum delta*, *Goniomya*, *Isocyprina*, *Modiolus*, *Myophorella*, *Nanogyra nana*, *Oxytoma*, *Pholadomya*, *Pinna sandsfootensis* and *Plagiostoma*, ammonites (*Ringsteadia* and rare *Amoeboceras*), gastropods (*Procerithium* and abundant *Dicroloma*), echinoid spines and a marine reptile (?opthalmosaur). Marine reptile bones, probably from the same bed, have been recorded from Gillingham brickpit (locality 7). These include the icthyopterygian *Macropterygius* sp. (Delair, 1960), the opthalmosaur *Opthalmosaurus pleydelli* (Delair, 1960; Mansell-Pleydell, 1890) and the plesiosaur *Pliosaurus macromerus* (Delair, 1959).

History of sedimentation

The sedimentation of the Corallian Group has long been described in terms of cyclic sequences. Arkell (1933) noted three clay-sandstone-limestone cycles in the Corallian of south Dorset. Other authors (Fürsich, 1976; Talbot, 1973; Wilson, 1968a; 1968b; Wright, 1986) have described a variety of cycles, mostly coarsening-upward. Sun (1989) reinterpreted the sequence as four transgressive–regressive cycles, each with an erosional base.

The Corallian Group in north Dorset shows an abrupt regression, from offshore shelf muds of the Oxford Clay, to subtidal clayey sands, followed by a series of minor, transgressive–regressive cycles, each culminating in a regressive, poorly sorted, intensely bioturbated, fine-grained and silty sand (Sands C and D in the **Hazelbury Bryan Formation**). These sands were probably deposited below the lower shoreface, and the intervening silty clay was deposited on a deeper water shelf.

The **Stour Formation** exhibits a series of coarsening-upward units with a transgressive ooidal clay at the base, passing up into ooidal limestone. The low sand content of the **Woodrow Clay**, compared with that of the Hazelbury Bryan Formation, may be due to reduction in supply caused by diminution of relief and an increase in the distance from the source. Laminations indicate gentle reworking, possibly on a tidal-flat; the bivalves suggest well-oxygenated waters.

The **Sturminster Pisolite** was probably deposited in a low-energy, possibly 'lagoonal' setting, although Wright (1981) refers to frequent agitation of the sediment. These cycles are similar to those in the Hazelbury Bryan Formation, except that the coarse-grained top consists of oolite or clayey oolite that probably represents sediment derived from shoals upslope, redeposited subtidally.

Sharply defined transgressions occur at the bases of the Newton Clay, Clavellata Beds, Sandsfoot Grit and Ringstead Waxy Clay. Cross-bedded, moderately sorted, very fine-grained sand or coarse silt at the top of the lower of two coarsening-upward sequences in the Newton Clay [7558 2997] suggest shoreface conditions.

The **Clavellata Beds** have a marked erosion surface at their base. They have a relative abundance of coarse shelly debris, consisting mainly of broken and abraded bivalves and gastropods. Sun (1989) suggested that they are lag deposits, related to a transgressive phase.

The **Sandsfoot Grit** has phosphatic nodules above an erosive base. It was probably deposited in the transition zone between offshore shelf and shoreface, or in the lower shoreface zone. The commonly ferruginous clay matrix suggests that it was deposited in an area of lower than

Figure 25 Outcrop of the Kimmeridge Clay (including Ringstead Waxy Clay and 'Ampthill Clay'): localities cited in the text and schematic section showing the stratal divisions and marker beds recognised within the district. Down to KC 27, the latter is based on Tisbury Borehole, to which the depths given for certain boundaries relate.

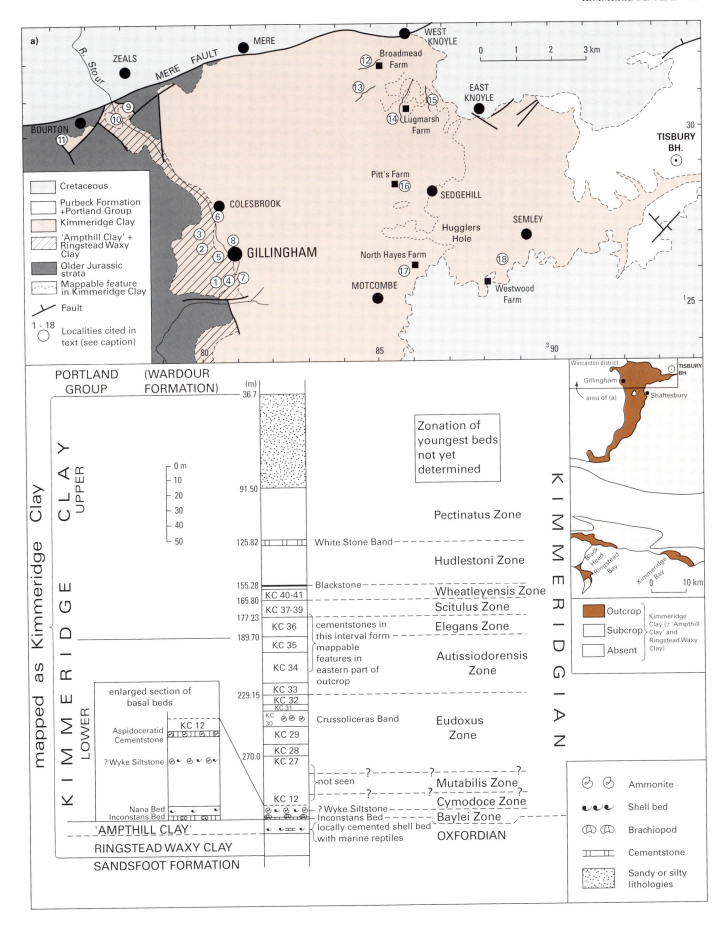

a)

ZEALS   MERE   WEST KNOYLE

⑫ Broadmead Farm

⑬   EAST KNOYLE

⑮

⑭ Lugmarsh Farm

TISBURY BH.

Pitt's Farm ⑯   SEDGEHILL

Hugglers Hole   SEMLEY

North Hayes Farm
⑰   ⑱

MOTCOMBE   Westwood Farm

COLESBROOK
⑥
③   ⑧
②   GILLINGHAM
⑤
① ④ ⑦

BOURTON ⑪   ⑩   ⑨   R. Stour   MERE FAULT

Cretaceous

Purbeck Formation +Portland Group

Kimmeridge Clay

'Ampthill Clay' + Ringstead Waxy Clay

Older Jurassic strata

Mappable feature in Kimmeridge Clay

Fault

1 - 18 ◯ Localities cited in text (see caption)

PORTLAND GROUP   (WARDOUR FORMATION)

mapped as Kimmeridge Clay

KIMMERIDGE CLAY

UPPER

LOWER

(m)
36.7

91.50

Zonation of youngest beds not yet determined

0 m
10
20
30
40
50

125.82   White Stone Band

155.28   Blackstone
165.80   KC 40-41
177.23   KC 37-39
KC 36   cementstones in this interval form mappable features in eastern part of outcrop
189.70   KC 35
KC 34
229.15   KC 33
KC 32
KC 31
KC 30   Crussoliceras Band
KC 29
KC 28
270.0   KC 27   not seen
KC 12

enlarged section of basal beds
KC 12
Aspidoceratid Cementstone
? Wyke Siltstone
Nana Bed
Inconstans Bed

'AMPTHILL CLAY'
RINGSTEAD WAXY CLAY
SANDSFOOT FORMATION

Pectinatus Zone

Hudlestoni Zone

Wheatleyensis Zone

Scitulus Zone

Elegans Zone

Autissiodorensis Zone

Eudoxus Zone

Mutabilis Zone

Cymodoce Zone

? Wyke Siltstone
Inconstans Bed
locally cemented shell bed with marine reptiles

Baylei Zone

OXFORDIAN

KIMMERIDGIAN

Wincanton district   TISBURY BH.
Gillingham
area of (a)   Shaftesbury

Black Head   Ringstead Bay   Kimmeridge Bay
0   10 km

Outcrop   Kimmeridge Clay (± 'Ampthill Clay' and Ringstead Waxy Clay)
Subcrop
Absent

Ammonite
Shell bed
Brachiopod
Cementstone
Sandy or silty lithologies

normal salinity, perhaps close to a delta or estuary. The
bivalves are consistent with such an environment. The
top of the member is marked by a further erosional break,
with phosphatised nodules and fossils in the sandy transgressive base of the **Ringstead Waxy Clay**. These sandy
beds pass up into relatively sand-free, pale grey clay with
an impoverished fauna. The sideritic nodules, possibly
indicative of lowered salinities, would explain the poor
fauna. The uppermost unit of the group, the **'Ampthill
Clay'**, is darker and contains a varied fauna more like
that of the Kimmeridge Clay, suggesting a more offshore
situation. Phosphatic nodules mark its transgressive base.

The north Dorset Corallian Group shows at least six
regressive–transgressive cycles. Some show an abrupt
transgressive base, while others are more gradational. On
the Cranborne–Fordingbridge High, south of the district,
the lower cycles and those in the Newton Clay are missing.
This lateral variability in cycle development suggests
local tectonic control on sedimentation, superimposed
on eustatic controls operating throughout the Wessex
Basin.

Thickness variation in the group shows evidence of
syndepositional fault control. The thickest deposits occur
in the Mere Basin; the group thins northwards towards
the Mendips, southwards over the Cranborne–Fordingbridge High and towards the south-east (Figure 23). In
the last area, the lowest part of the Hazelbury Bryan Formation and the Sandsfoot Formation are absent, probably
due to movements on a north-west-trending fault.

Kimmeridge Clay (Formation)

The outcrop of the Kimmeridge Clay is restricted to the
south of the Mere Fault. The complete succession is only
preserved in the east where the formation is overlain by
the Portland Group. Elsewhere, the formation is variously
truncated beneath the basal Cretaceous unconformity. A
tentative position for its lower boundary, based largely on
palaeontology, is shown on Figure 25.

Comparison with sections in the adjoining Shaftesbury
district suggests that the Lower Kimmeridge Clay is about
120 m thick; the Upper Kimmeridge Clay is about 160 m
thick.

The Kimmeridge Clay mudstones are, at various levels,
calcareous, kerogen-rich (bituminous mudstones and oil
shales), silty or sandy. They occur in a complex sequence
of small-scale rhythms, and are associated with thin siltstone and cementstone beds. At its type locality on the
Dorset coast, the formation is divided into Lower and
Upper Kimmeridge Clay which are readily distinguished
on the basis of ammonite genera (Cox and Gallois, 1981).
The base of the formation is defined by the **Inconstans
Bed**. Although this bed is present in the district, the
eponymous brachiopod *Torquirhynchia inconstans* is much
rarer than on the Dorset coast. It was recorded by Woodward (1895) from the lower part of Gillingham brickpit
(locality 7) and, together with *Pictonia*, in sections at and
near the Zeals underpass (locality 9) during construction
of the new A303 (Bristow et al., 1992). Farther west, in
the cutting near the River Stour (locality 10), the Incon-

stans Bed yielded thecosmiliid corals [7792 3035]; *T.
inconstans* and *Pictonia?* were recorded [7617 3004] in a
cutting 8 m deep at Chaffeymoor Farm (locality 11).

Other marker beds of the coastal sections (Cox and
Gallois, 1981) have been noted in this district. The **Nana
Bed** was found at Colesbrook (locality 6), where a spoil
heap [8065 2780] showed clay with *Nanogyra nana*, and a
piece of a cemented lumachelle composed of *N. nana*
valves with *Grammatodon?*, *Modiolus?*, serpulids and an
echinoid spine. The Nana Bed was probably also exposed
in the Gillingham brickpit (locality 7) (Woodward, 1895).
The **Wyke Siltstone, Black Head Siltstone** and a '*Xenostephanus*'-rich horizon also occur in this region; the last-
named was noted in the Chaffeymoor Farm cutting
(locality 11) where crushed bivalves and ammonites found
loose [around 764 300] include *Liostrea*, *Thracia* and
Rasenia ex gr. *evoluta*, including *Xenostephanus*-like forms.
A weathered representative of the Wyke Siltstone may
have occurred at the top of the Gillingham brickpit
(locality 7) ('a shelly layer with large ammonites
[raseniids]…too decayed to be brought away'). In addition,
a new marker bed of cementstones with large aspidoceratid ammonites, the **Aspidoceratid Cementstone**, was
recorded near the Zeals Underpass (locality 9), Stour
cutting (locality 10), on the Gillingham relief road (locality
5), and at the old Bay swimming pool (locality 8) (Bristow,
1993; Bristow et al., 1992). The aspidoceratid compares
with *Paraspidoceras rupellense* morph. *compressum* of Hantzpergue (1989), and is associated with an early *Rasenia* of
the Cymodoce Zone. In France, this *Paraspidoceras* char-
acterises the Cymodoce and Achilles subzones, that is the
middle part of their Cymodoce Zone. This suggests that
the Aspidoceratid Cementstone occurs at about the level
of the Black Head Siltstone.

The Kimmeridge Clay has been divided into numbered,
small-scale, stratigraphical units (hereinafter referred to
as KC 1, KC 2 etc.) (Cox and Gallois, 1979; 1981; Gallois
and Cox, 1976). This scheme, based on faunal markers
and variation in lithology, is shown, together with the
ammonite-based chronostratigraphy, in Figure 25.

During this survey, exposures in four streams (Figure
25, localities 12, 13, 14, 15) south of West Knoyle proved
KC 29–KC 35 in the upper part of the Lower Kimmeridge
Clay (Barton, 1994; Bristow, 1995a). Moderately fissile,
medium and pale grey, variably shelly mudstones and oil
shales yielded an ammonite fauna of the Eudoxus and
Autissiodorensis zones, including species of *Amoeboceras
(Nannocardioceras)*, *Aspidoceras* (and *Laevaptychus*), *Aulacostephanus* and *Sutneria*, as well as a typical assemblage of
bivalves, gastropods, scaphopods and *Lingula*. In particular, the occurrence of *Sutneria rebholzi* [8482 3110] (locality
13) indicates KC 33, and a worn large *?Aulacostephanus
autissiodorensis* [8579 3037] (locality 14) almost certainly
came from KC 34. Farther upstream, ammonite fragments
including pectinatitids, and possible *Subdichotomoceras
websteri* as well as *Aulacostephanus*, indicate the highest part of the Lower Kimmeridge Clay (KC 34 and
KC 35).

These stream sections also throw light on the stratigraphical position of three mappable features which occur

hereabouts (Figure 25). The lowest [8657 3085] (locality 15) is a bed of markedly septarian concretions; its associated fauna (Bristow, 1995a), and comparison with the coastal sections at Black Head and Ringstead Bay (Cox and Gallois, 1981), suggest that it lies in the basal part of KC 35 or top KC 34. The second feature-forming bed is a densely cemented, brownish grey cementstone which forms a small waterfall [8608 3122] (locality 14), but its stratigraphical position appears to be anomalous unless there is some undetected structural complication. The fauna from the overlying oil shale includes *Nanogyra virgula*, clusters of *Nicaniella extensa*, abundant *Protocardia*, pectinatitid ammonite fragments and clusters of the small trochiform gastropod *Semisolarium hallami*. Elsewhere, this faunal assemblage characterizes KC 37 in the lower part of the Upper Kimmeridge Clay. However, this interpretation conflicts with occurrences of *Aulacostephanus* (which does not range above KC 35) both upstream and downstream. The highest feature-forming bed must lie in KC 35 or the lowest beds of the Upper Kimmeridge Clay, and certainly close to their mutual boundary. An exposure [8613 3137] (locality 14), not far below the feature, yielded *Aulacostephanus*, but there is no faunal control from above it. The feature has been traced into the Sedgehill area as far south as the Cretaceous unconformity south of Huggler's Hole. There, it is estimated to occur 25–30 m above a distinctive 10 to 20 mm-thick chalky layer, composed mainly of the ammonite *Subdichotomoceras websteri* with a few *Aulacostephanus*, which was exposed in a slurry pit [8563 2848] (locality 16) at Pitts Farm. This bed belongs in the Autissiodorensis Zone, probably high KC 34 or low KC 35. Assuming stratal thicknesses are comparable with those in the Tisbury Borehole [9359 2907], which cored the Kimmeridge Clay (down to KC 27) (Bristow, 1995b; Gallois, 1979), the feature is thought to lie at about the level of the Yellow Ledge Stone Band (basal KC 37) in the lower part of the Upper Kimmeridge Clay. If it is indeed the same feature as that developed in the West Knoyle area and if there is no undetected structural complication there, then a slightly lower cementstone, at about the level of Blake's Bed 42 and the Lower–Upper Kimmeridge Clay boundary, is more likely.

There is some uncertainty about the precise stratigraphical position of fissile, brownish and medium grey mudstone exposed in stream sections [8578 2600] (locality 17) south-west of North Hayes Farm, Motcombe. The fauna at all three localities is dominated by bivalves, including clusters of *Nicaniella extensa*, as well as pectinatitid ammonite fragments. The lithology and fauna suggest a position low in the Upper Kimmeridge Clay (KC 36 or KC 37) or high in the Lower Kimmeridge Clay (KC 35). Palynological dating (Riding, 1993b, unpublished report*) indicates the Upper Kimmeridge Clay, but maybe as high as KC 40. Assuming no structural complication, the topographical position relative to the mappable feature near Sedgehill suggests that the exposures are more likely to be within the Lower Kimmeridge Clay. A septarian cementstone nodule with a broken, but uncrushed, large *Pectinatites* indicative of the Upper Kimmeridge Clay, was collected from a ditch [8857 2612] (locality 18) at Westwood Farm, Semley.

PORTLAND GROUP

The Portland Group comprises the Wardour Formation overlain by the Portland Stone Formation.

The Wardour Formation corresponds to the Lower Portland Beds of Hudleston (1881), Woodward (1895), Reid (1903) and others, and to the Wardour Member of Wimbledon (1976) (Table 10). The latter term was introduced for the basal, dominantly sandy, unit of Wimbledon's Portland Sand Formation, the other members of which (Chicksgrove and Tisbury) are mainly limestones. On lithostratigraphical grounds, these limestones are more appropriately grouped in the Portland Stone Formation, and, alone, the Wardour Member of Wimbledon's sequence is given formational status (Table 10).

The Portland Stone Formation of this account corresponds to the Upper Portland Beds of Woodward (1895). It is divided into three members, in ascending sequence, the Tisbury, Wockley and Chilmark members (Bristow, 1995b). The member names were introduced by Wimbledon (1976), but were modified by Bristow (1995b) (Table 10). In mapping, it was not possible to separate Wimbledon's Chicksgrove and Tisbury members. Accordingly, the Tisbury Member as used in this account includes the Chicksgrove Member as well as the Ragstone, included by Wimbledon (1976) at the base of the overlying Wockley Member.

Wardour Formation

The formational name was introduced by Wimbledon (1976). The type area is the Vale of Wardour. No type section was designated, but it is presumed to be the Chicksgrove Quarry [962 296] just east of the district.

The Wardour Formation crops out in the south-east of the district. It is 23.7 m thick in the Tisbury Borehole, but at outcrop, the thickness ranges from 10 to 15 m, maybe as a result of cambering of the overlying limestones over part of the crop.

The base of the formation is taken at the base of a fine-grained, glauconitic sandstone, such as seen south-east of Ashleywood [9372 3073], where very fine-grained orange sand is thrown out from burrows. It has also been recorded south of Tuckingmill, where Mottram (unpublished map, BGS) noted 'buff sand with a rubble of an *Exogyra* band' [9352 2881], near Wardour [9231 2707, 9236 2674, 9248 2671 and 9265 2663]. The gamma-ray log of the Tisbury Borehole [9399 2907] shows this sandstone to be part of a coarsening-upward cycle (Figure 26). Springs issue from the base of the sandstone and enable the base of the formation to be traced with ease in the field.

The bulk of the formation consists of siltstone and fine-grained, bioturbated, friable, sparsely shelly sandstone (Plate 8h). In the Tisbury Borehole, lydite (a hard, silicified shale) pebbles (3–4 mm across) were noted 5.13 m above the base. Rare, brown and black lydite pebbles (up to 4 mm across) have also been noted in the Nadder Valley [9180 2528]. Wimbledon (1976) recorded a lydite bed 3.5 m from the top of the formation at Chicksgrove, just east of the district. A borehole [9295 2666] in Wardour Park proved 1.7 m of greyish brown mottled, clayey silt,

Table 10 Stratigraphical nomenclature of the Portland Group of the district.

Blake (1881)	Woodward (1895)			Wimbledon (1976)		Wincanton district
				Members		
Building Stone Series *	UPPER PORTLAND BEDS	Upper Building Stones *		PORTLAND STONE FORMATION — Chilmark		PORTLAND STONE FORMATION — Chilmark
Chalk of Chicksgrove		Chalky Series		Wockley		Wockley
Upper beds of Swindon		Ragstone				
Tisbury Freestone *		Lower Building Stones	Trough Bed	PORTLAND SAND FORMATION — Tisbury		Tisbury
			Glauconitic & sandy limestones †	Chicksgrove		Sand
Lower Sands	PORTLAND SAND FORMATION	Lower Portland Beds		Wardour		Wardour Formation

* Main building stones

† Divisible in descending sequence into Green Bed, Slant Bed, Pinney Bed, Cleaving or Hard Bed, Fretting Bed and Under Beds

on 4.7 m of orange-brown and grey, silty, fine- to medium-grained sand. Sandy mudstone from the inlier [9180 2530], beneath drift deposits, in the Nadder valley at Donhead St Andrew, yielded long-ranging Portlandian foraminifera (Wilkinson, 1993d, unpublished report*).

Dinoflagellate cysts indicate that the basal 5 m of the Wardour Formation, a coarsening-upwards, glauconitic sandstone, fall in the youngest zones of the Kimmeridgian Stage. The succeeding part spans the Albani Zone and basal part of the Glaucolithus Zone (Cope et al., 1980b; Riding, 1993c, unpublished report*; Wilkinson, 1993b, unpublished report*).

Portland Stone Formation

The lower part of the Portland Stone Formation consists dominantly of micrite, with subordinate siltstones, fine-grained calcareous sandstone and sandy limestone. It is part of the Chicksgrove Member of Wimbledon (1976). This passes up into fine-grained sandy, glauconitic, peloidal, biosparites, the Tisbury Member (Wimbledon, 1976), overlain by a peloidal, bioclastic micrite, the Ragstone (Woodward, 1895). The above sequence comprises the Tisbury Member of Bristow (1995b). The Tisbury Member is overlain by shelly, micritic limestone with little or no siliciclastic sand, the Wockley Member (Wimbledon, 1976). In this district, the youngest unit, the Chilmark Member, consisting of fine-grained oolites, occurs only in the area south of Fonthill Bishop [925 316, 932 303 and 942 314].

TISBURY MEMBER

The name was introduced by Wimbledon (1976). No type section was specified, but by implication, it is the Chicksgrove Quarry east of the district. The member corresponds mainly to the Lower Building Stones of the older literature, but, as defined herein, also includes the Ragstone (Table 10). Its base is taken at an abrupt lithological change, from fine-grained, clayey sand of the Wardour Formation to calcareous sandstones and sandy limestones; this coincides with a marked positive topographical feature.

The member consists of micrites and siliciclastic and peloidal, bioclastic, glauconitic biosparites (Plate 8c–g). Chert occurs sparingly in the member. Lydite pebbles were noted in the basal bed at Hazeldon Farm [around 9376 2816] (Hudleston, 1881). The glauconite content varies from 1 to 2 per cent, the bioclastic sand up to 31 per cent (average about 8 per cent), siliciclastic sand from 4 to 42 per cent (average about 30 per cent), and peloids from 5 to 26 per cent (average about 10 per cent). There is a general gradual upward increase in bioclastic sand and a decrease in siliciclastic sand, and the beds pass up into bioclastic biosparites. The Tisbury Freestone of Blake (1880), the principal source of building stone in the district, corresponds to this higher part of the Tisbury Member; it has a wide outcrop north of the River Nadder.

In places, such as south-east of Hatch House [around 915 279], and between Hazeldon Farm and Wardour Castle [927 270], a three-fold division of the Tisbury

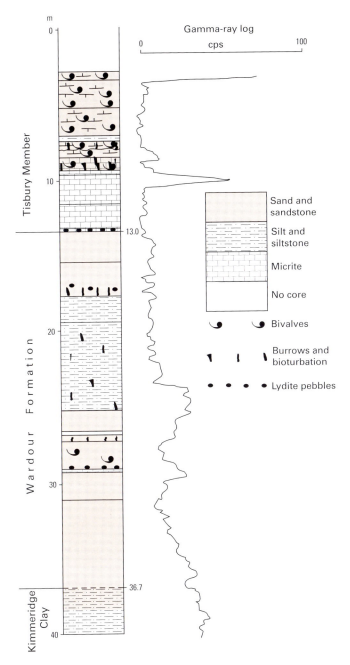

Figure 26 Lithostratigraphy of the Portland Group in Tisbury Borehole [9399 2907].

Member can be mapped. In upward succession, this consists of up to 5 m of shelly, generally micritic limestone (possibly the Chicksgrove Member of Wimbledon (1976)): up to 5 m of buff, silty, fine-grained, glauconitic sand and, locally, khaki clay: peloidal, sandy, bioclastic, sparry limestones with common *Myophorella* and ammonites. Where the median sand is absent, or cannot be recognised on pasture-covered ground and steep valley sides, such as south of Tisbury, no firm boundary can be drawn between the lower and higher limestones.

A shelly micrite, including *Myophorella*, is common at the base of the member from just east of East Hatch to

Tisbury [9328 2850 to 9345 2882, and 9357 2885 to 9374 2863 to 9405 2865], and on the sides of Oddford Brook valley [9395 2925 to 9367 2942, and 9345 2965 to 9373 2966]. Hudleston (1881) recorded a section [around 9376 2816] in a lane near 'Hazleton' through the lower limestone (0.9 m of a shelly 'greenish, concretionary limestone grit, with occasional lydite') and overlying sand (2.1 m of 'loose sand with doggers') units.

Tisbury Quarry [931 291], the last quarry to close in the district, was worked in the middle and upper part of the member. A representative section [9320 2904] exposes (Plate 3):

	Thickness m
Limestone rubble	0.7
Limestone, sparry, bioclastic (24%), fine-grained siliciclastic (22%), sparsely glauconitic (2%), peloidal (27%)	0.9
Biosparite, fine-grained sandy, peloidal, massive, passing laterally into a flaggy biosparite; at base, lenticular chert nodules up to 0.7 m long by 0.08 m thick	0.8
Limestone, sparry, bioclastic (32%), fine-grained siliciclastic (21%), peloidal (18%), sparsely glauconitic (4%) porous (22%); passes down into less bioclastic (21%) and sandy (10%), peloidal (5%), glauconitic (2%), very porous (28%) sparry limestone in beds 0.3 to 1.5 m thick. Some bedding surfaces covered with large bivalves	3.2

Two dominant joint directions trend at 130° and between 190° and 205°. An illustrated account of the working of this pit is given by Manners (1971). Another typical section is seen in a partially overgrown pit [9351 3107] at Fonthill Gifford, where 5.4 m of the Tisbury Member is exposed:

	Thickness m
Sandstone, bioclastic, massive, weathering flaggy	0.70
Sandstone, as above, massive, but splitting irregularly; coarse-grained bioclastic sandstone, 0.15 m thick at base	0.85
Sandstone, hard, bioclastic shelly sandstone with fossils as hollow moulds ('Roach')	0.6–0.8
Sandstone, hard, glauconitic, bioclastic, in three principal beds	0.75
Chert, tabular, irregular	0.05
Sandstone, bioclastic, planar bedded at top	0.3–0.7
Chert, passing into dark brown, silicified, cross-bedded sandstone	0.25
Sandstone, friable to hard, bioclastic, glauconitic in beds 0.1 to 0.5 m thick; lenticular chert (up to 0.1 m thick) over a 4 m length	0.95

The valleyward dip of the beds, up to 40°, is presumed to be a cambering feature.

A quarry [9400 2809] 500 m east-north-east of Hazeldon Farm exposes the highest part of the member:

	Thickness m
Oobiosparite	1.10
Chert, tabular, locally absent, in which case a marked recess occurs in the face	0 to 0.07

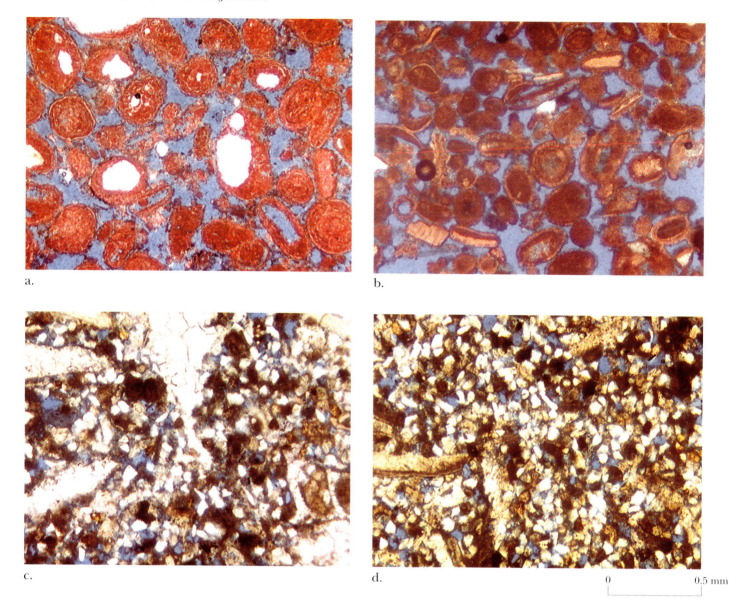

0   0.5 mm

Plate 8 Photomicrographs of limestones and sandstones of the Portland Group (aH plane polars, blue resin-impregnated and dual-carbonate stain)

a. **(VR34) Chilmark Member**   Non-ferroan calcitic (pink stained) micritised, medium to coarse sand-grade, ooidal grains forming a porous (blue-stained pores) very weak framework. Ooid nuclei range from abraded bioclastic debris to monocrystalline quartz (white) grains. A narrow fringing microspar cement coats many grains.

b. **(VR35) Chilmark Member**   Non-ferroan calcitic (pink), poorly sorted, micritised, medium to coarse sand-grade ooid and peloidal grains forming a porous (blue-stained pores) very weak framework. Ooid nuclei are commonly abraded bioclastic fragments (pale pink) or rarely siliciclastic grains (white).

Tisbury Borehole

c. 3.25 m **Tisbury Member**   Very fine-grained, porous, bioclastic sandstone comprising well-sorted, very fine monocrystalline quartz, with subordinate feldspar. Very coarse bioclasts commonly occur, many are at least partially replaced by coarse, non-ferroan calcite spar or chalcedonitic (siliceous) cements. Intergranular micritic muddy matrix commonly disrupts the porosity.

d. 6.25 m **Tisbury Member**   Very fine-grained, porous sandstone, dominated by monocrystalline quartz with subordinate potassium feldspar and fine bioclastic debris. Sparse coarse, non-ferroan calcite bioclastic debris and rare glauconite grains are present. A patchy intergranular micritic matrix is present.

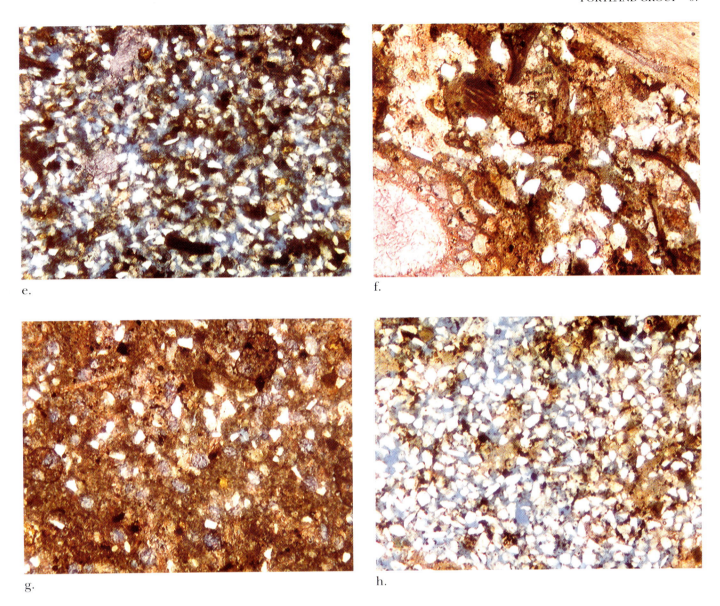

e.

f.

g.

h.

Plate 8   continued.

e. 7.35 m **Tisbury Member**   Porous, micritic, bioclastic, very fine-grained sandstone/coarse siltstone, dominated by monocrystalline quartz with subordinate potassium feldspar in a micritic matrix. The bioclastic debris is commonly replaced by ferroan (purple stained) spar calcite.

f. 9.60 m **Tisbury Member**   Sandy biopelmicrite with abundant fine and coarse bioclastic debris (bivalve, bryozoan). The bioclasts are either ferroan calcite spar-replaced or are non-ferroan calcite and retain their original wall structure. The micritic 'matrix' has a distinctly peloidal texture in places. Some bioclasts have extensive siliceous (chalcedonitic) patches. Very fine sand-grade siliciclastic grains are abundant. Poor porosity.

g. 10.35 m **Tisbury Member;** *'Spicular' micrite.*   Abundant ferroan spar-replaced spheres (purple stained) and non-ferroan calcite bioclastic debris, of fine sand-grade, in a micritic, non-ferroan calcite matrix. The micritic matrix is vaguely peloidal in some patches. Concentrations of the sparry carbonate spheres (original siliceous sponge spicules?) may coalesce into extensive ferroan spar cement patches.

h. 14.1 m **Wardour Formation**.   Laminated, very fine-grained, porous, muddy micritic sandstone dominated by monocrystalline very fine quartz grains. No potassium feldspar grains apparent. Sparsely glauconitic with patchy, pyritic, muddy matrix and micritic patches.

	Thickness m
Biosparite, fine-grained sandy, weakly glauconitic, lenticular chert nodules up to 0.05 thick by 0.45 m long	1.00
Biosparite as above; cross-bedded in lowest 0.20 m	0.50 to 0.75
Biosparite, fine-grained sandy, weakly glauconitic, cross-bedded, in four beds	2.25

Fossils, especially bivalves (including *Myophorella*) and ammonites, are common in the Tisbury Member. The coral *Isastraea oblonga* (known as the Tisbury Star) occurs locally west [935 310] and south [936 307; 9376 3028] of Ashleywood. Ammonites from the lower part of the member at Chicksgrove and Chilmark, just east of the district, include common *Glaucolithites glaucolithus* and *G. polygyralis* indicative of the Glaucolithus Zone (Wimbledon, 1976). Ammonites in the upper part, where glaucolithitids are absent, include species of *Titanites (Titanites)*, *T. (Polymegalites)* and *Galbanites* indicative of the Okusensis Zone (Wimbledon, 1976; 1980).

WOCKLEY MEMBER

The term Wockley Member was introduced by Wimbledon (1976). It corresponds to the Chalky Series of Woodward (1895). The name is believed to be a Wiltshire dialect corruption of 'Oakley' (Edmunds, 1938). No type section was specified by Wimbledon, but by implication, it is the Chicksgrove and Chilmark quarries east of the district. The former Oakley (or Wockley or Shavers Bridge) Quarry [9555 2870] of Andrews and Jukes-Browne (1894) may be regarded as a reference section.

The Wockley Member crops out on either side of the southward-flowing stream from Fonthill Bishop [933 303 and 943 310] and around Newtown [920 290]. North of Tisbury, the member is cut out northwards beneath the Chilmark Member. South of Tisbury, the outcrop cannot be traced across pasture-covered ground, but one pit [943 282] formerly exposed 'Chalky Portlandian' (Mottram, unpublished map, BGS).

Wimbledon (1976) drew the base of the Wockley Member at the base of the 'Ragstone'. This unit, and especially its basal shelly micrite, is readily recognisable in quarry sections, but not always easily identifiable from brash. At Chicksgrove, just east of the district, the basal micrite is 0.5 to 0.85 m thick and very hard with common large bivalves. It is succeeded by up to 2.5 m of micritised, peloidal limestone with common bivalves and scattered large ammonites; these limestones are lithologically more like the underlying Tisbury Member than the overlying part of the Wockley Member, particularly where seen only as brash, and have been mapped with the former.

In the field, the base is taken at the incoming of dense, porcellanous, grey limestone with moulds of common small gastropods. This is probably the 'hard, white-cream lime mud with abundant minute gastropods' near the base of the Wockley Member in the Chicksgrove Quarry (Wimbledon, 1976). Higher in the sequence, chalky, shelly micrites with common chert occur and typify the member.

The ammonites *Galbanites (Kerberites) kerberus*, *Titanites giganteus*, *T. trophon* and *T.* cf. *anguiformis* have been found in the Wockley Member and indicate the Kerberus Zone and the lowest part of the Anguiformis Zone (Wimbledon, 1976; 1980).

CHILMARK MEMBER

The Chilmark Member, the Upper Building Stones of the older literature, was thought to be restricted to the Chilmark ravine, east of the district. However, this survey shows that it extends westwards into the Fonthill Gifford area [925 315 and 932 305], where it typically forms a brash of fine-grained, flaggy oolite (Plate 8a–b). The outcrop on the east side of Fonthill Lake also includes beds of coarse-grained, ooidal, bioclastic sand, locally with small gastropods. The gastropod *Aptyxiella portlandica*, the Portland Screw, is common in this member (Wimbledon, 1976). The junction with the overlying Purbeck Formation is transitional, with interbedded stromatolitic tufas and ooidal sands forming a transitional zone.

The Chilmark Member was widely quarried as a building stone in the type area, but it has not been worked in this district.

No ammonite has been found in the Chilmark Member, but Wimbledon (1976) and Cope et al. (1980b) regarded the Wockley and Chilmark members largely as lateral equivalents, with the latter falling in the upper part of the Kerberus Zone and the lowest part of the Anguiformis Zone.

Purbeck Formation

OAKLEY MARL MEMBER

Only strata equivalent to the lower part of the Lower Purbeck Beds of the traditional classification (Woodward, 1895) crop out in the district. They have been named the Oakley Marl Member (Bristow, 1995b), with the type locality being the old Oakley (or Wockley or Shavers Bridge) Quarry [9555 2870] south-east of Tisbury (Andrews and Jukes-Browne, 1894).

The member crops out south of Fonthill House [around 944 317], in a fault-bounded crop [937 319] west of Fonthill, and south of Fonthill Gifford [9250 3145], there is no exposure. For mapping purposes, the base of the formation is taken at the incoming of marl and marly limestone, above limestones of the Portland Group. This may exclude some thin beds of limestone which are classified with the Purbeck Formation in quarry sections (Woodward, 1895).

The Oakley Marl Member consists of at least 5 m of buff and dark grey marly clay, with thin beds of marly limestone.

SEVEN

Cretaceous

Cretaceous rocks crop out principally in the north-east of the district, but there is a small outcrop in the extreme south-east. The Lower Cretaceous rocks in the district comprise, in ascending order, the Lower Greensand, Gault and Upper Greensand formations.

LOWER GREENSAND (FORMATION)

The formation probably corresponds to the Child Okeford Sands of the Shaftesbury district (Bristow et al., 1995). The overlying Bedchester Sands are either absent or have been included with the Gault (see below).

Lower Greensand, up to 5 m thick, crops out on either side of the River Nadder. The main occurrences are south of Fonthill Gifford [932 314 and 931 308], south of Fonthill House [944 319] and an almost continuous outcrop from just east of Gutch Common [900 258] to the margin of the district.

There is no exposure, but orange-brown or glauconitic sand can be augered over most of the crop. It is fine to medium grained, locally clayey and, in part, well sorted. Locally [9166 2501 and 9476 3151], poorly sorted, coarse silt occurs. South-west of Fonthill House, ferruginous, fine-grained sandstone forms a northward-dipping slope [9429 3185 to 9463 3174]. Mottram (1961) recorded small, rounded quartz pebbles in sand in a pit east of the district, and refers to the ease with which the formation can be mapped by the scatter of polished pebbles across the outcrop. However, as no pebble has been found in augering, it is probable that most of these surface pebbles are derived from the basal bed of the Gault. The base of the formation is commonly marked by springs [e.g. 9041 2620 to 9074 2642 and 9091 2639 to 9088 2627].

No diagnostic fauna has been found in the Lower Greensand of the district, but, as elsewhere, it is presumed to be of Aptian and Early Albian age. Reid (1903) recorded 'cherty sandstone with '*Pecten (Neithea) quinquecostatus*' and '*P. orbicularis*' from just east of the district [around 9485 2820]. The sparse microfauna from glauconitic fine-grained sand, beneath head deposits in a trench [9166 2501] south-east of Beauchamp House, includes *Tritaxia* sp., *Arenobulimina* cf. *chapmani* and *A. macfadyeni* indicative of an Early to Mid-Albian age.

GAULT (FORMATION)

In the Shaftesbury district, the Bedchester Sands at the top of the Lower Greensand consist of very clayey, very fine-grained sand. In the absence of the basal Gault pebble bed, or in areas where it has not been recognised, it is probable that the Bedchester Sands have been included with the Gault (Bristow et al., 1995). This may be the situation in the south-west of the district around, Bartholomew Hill [905 256], and on Round Hill [915 260], where a bed, up to 5 m thick of clayey, glauconitic, fine-grained sand lies about 7 m above the mapped base of the Gault. Support for this interpretation is provided by the spore *Polypodiaceoisporites foveolatus* from a clayey sand about 10 m above the mapped base of the Gault [9032 2590]; this falls in the range Early Aptian (*deshayesi* Zone) to Early Albian and is characteristic of the Lower Greensand of southern England.

North of the Mere Fault, the Gault crops out in an arc from just north of Horningsham to just west of Bourton, with an outlier on Postlebury Hill [735 428]. South of the fault, the Gault crops out on either side of the Vale of Wardour. On the north side of the vale, the steep dips of the Wardour Monocline give rise to a very narrow outcrop. Much of the outcrop is obscured by landslip.

The formation is between 10 and 25 m thick. It consists dominantly of glauconitic, micaceous, fine-grained sandy clay. The base is locally marked by springs [e.g. 9010 2626]. Small, well-rounded pebbles are common in the basal bed over much of the outcrop. On Postlebury Hill [735 428], the basal bed, up to 0.4 m thick, consists of a pebbly, gritty sand to sandy clay, with scattered, centimetre-sized, pebbles of quartz and 'lydite' (red siliceous material). The pebbly basal bed was found at East Knoyle [c.8611 3259] (Jukes-Browne, unpublished, BGS manuscript), south-south-west [9278 3033 to 929 308] and south west [9264 3117 and 9229 3113] of Fonthill Gifford church. The basal beds of the Gault were exposed in a pit [9343 2673] north-west of Ark Farm, where Lower Greensand is overlain by 0.35 m of blue clay with small quartz pebbles towards the base, followed by 0.4 to 0.5 m of a nodular ferruginous layer, and capped by 0.9 m of blue clay with nodules (Mottram, 1957). Elsewhere, the base of the Gault may be a clayey, fine-grained, glauconitic sand, for example south-west of Broad Oak [900 263] where it consists of very sandy, glauconitic clay and clayey sand. On the opposite side of the valley, glauconitic, fine-grained sand occurs in the lower part of the Gault, as proved in auger holes near Hatt Farm [8772 2565] and just west of Westwood Farm [8806 2561 and 8822 2572].

In the north of the district, boreholes [around 840 425] north-west of Shear Water proved more than 19 m of firm to stiff, dark grey, very silty clay, interlayered with very silty sand. A well [8758 3047] at East Knoyle proved (Jukes-Browne, manuscript, BGS):

	Thickness m
Soil	0.6
Gault	
Clay, brown and yellow	2.4
Clay, dark grey silty	6.1
Sand, dark green glauconitic	about 6.1
Kimmeridge Clay	
Clay, dark grey	6.1

In spoil from this borehole, Jukes-Browne found several small phosphatic nodules and fragments of ammonites, including '*Ammonites splendens*'. The glauconitic sand may be a local development of Lower Greensand, but augering elsewhere in this area has not proved such a thickness.

The only good exposure in the Gault was that formerly seen in the Crockerton brickpit [868 423] (Ponsford, manuscript, BGS), where a ferruginous siltstone shows that the Gault is folded into steeply dipping (dips up to 50°) anticlines and synclines with axes parallel to the Wylye valley. These folds are presumably a result of valley bulging. Jukes-Browne and Hill (1900) recorded 8 m of dark grey clay with phosphatic septarian nodules (7 to 8 cm across), overlying a layer of lenticular masses (15 to 20 cm across) of hard, dark grey sandy stone. Many fossils have been recorded from this pit (Jukes-Browne and Hill, 1900; Woods and Bristow, 1995), which was the source of the holotypes of '*Ammonites bennettianus,* and, *A. laevigatus*', and probably the '*Ammonites monile*' mentioned by Fitton (1836), as well as many of the *lyelli* Subzone ammonites used by Spath (1923–1943) in his monograph of the Gault Ammonoidea (Owen, 1971). Amongst the fossils, and presumably from the basal beds, were specimens of *Lyelliceras lyelli*. Other *lyelli* Subzone fossils appear to have come from the lower part of the upper unit, with the higher part yielding *spathi* Subzone ammonites (Owen, 1971). Other specimens include *Hoplites* ex gr. *benettianus, H. dentatus, H. paronai* and *H. spathi*, indicative of the Middle Albian *lyelli* and *spathi* subzones of the *dentatus* Zone.

UPPER GREENSAND (FORMATION)

Jukes-Browne and Hill (1900, pp.157, 235) made a five-fold subdivision of the Upper Greensand near Warminster and Shaftesbury (Table 11). The lower two units were grouped as the Devizes Beds (or Zone of *Ammonites rostratus*), and the three upper units as the Warminster Beds (or Zone of *Pecten asper* and *Cardiaster fossarius*).

The Upper Greensand has a wide outcrop across the district, especially north of the Mere Fault. The outcrop is mostly fairly narrow on the north side of the Vale of Wardour, where the outcrop is affected by the Wardour Monocline. Much of the outcrop of the lowest member (Cann Sand) is landslipped. The base is commonly marked by springs.

The member names used in this district were introduced by Bristow (1989a). The formation is about 70 to 75 m thick in the north of the district around Maiden Bradley and Horningsham, and thins eastwards; 32 m were proved in a borehole [869 404] at Hill Deverill, and just north of the district, boreholes east of Warminster [around 927 447] proved between 15 and 18 m of Upper Greensand. A similar marked thinning is seen at outcrop north of Warminster on Sheet 281 Frome. Southwards, about 70 m was proved in the Norton Ferris Borehole [7820 3700] and it is at least 65 m thick in the Alfreds Tower area. South of the Mere Fault, the Upper Greensand is about 55 m in the Donhead area, and 30 m near Berwick St John, just south-east of the district.

Table 11
Classification of the Upper Greensand of the Wincanton district and Warminster area.

Wincanton district		Jukes-Browne and Hill (1900)		
	Thickness m			Thickness m
Melbury Sandstone	1–4	Warminster Beds	Greensand, fossiliferous with nodules and layers of calcareous stone	c.1.2–3
Boyne Hollow Chert	8–20		Chert beds and sands and sandstone	7.5
Shaftesbury Sandstone (Ragstone at top)	15–25		Greensands with layers of glauconitic limestone or greensand-rock	2.1–3.6
		Devizes Beds	Green, grey and buff sands more or less micaceous, with '*Exogyra conica*' and passing down into soft micaceous sandstone with large 'burrstones'	21–30
Cann Sand	5–18		Pale grey malmstone (Malmstone)	6

The lithology of the Upper Greensand of Wessex has been described by Tresise (1960; 1961); he included few observations from the Warminster district, but his general conclusions are relevant.

There is little faunal evidence for the age of most of the Upper Greensand, except the Melbury Sandstone. Few ammonites have been found, and for some, the exact stratigraphical provenance is uncertain. However, by combining the sparse ammonite evidence from western Wiltshire with that from north Dorset (Bristow et al., 1995), a broad chronostratigraphical framework can be established (see below). The Upper Greensand ranges from the Mid Albian *inflatum* Zone, *varicosum* Subzone, to the Early Cenomanian *dixoni* Zone. The detailed fauna is listed in Woods and Bristow (1995).

Cann Sand (Member)

The Cann Sand equates with the 'Malmstone' of Jukes-Browne and Hill (1900) — a fine-grained siliceous rock, with the silica of a 'colloidal variety'. The type area is the village of Cann [872 213], south of Shaftesbury (Bristow, 1989a). The outcrop occupies a shelf-like feature below the scarp formed by the Shaftesbury Sandstone. Many outcrops are involved with landslips and cambers. For example, most of Donhead St Mary in the south-east is built on free-draining, relatively undisturbed, but cambered sands, whilst the surrounding farm land is on poorly draining, hummocky, landslipped Gault and Cann Sand. The lower boundary with the Gault is sharp and commonly marked by springs. The upper boundary with the Shaftesbury Sandstone is usually marked by a sharp negative feature.

North of the Mere Fault, the Cann Sand, is between 15 and 18 m thick, and between 5 and 15 m thick around the Donheads in the south.

The Cann Sand is dominantly a glauconitic, poorly sorted, very sandy silt to very fine-grained, micaceous sand. Much of the member is uncemented, but Jukes-Browne and Hill (1900) noted 2.4 m of pale grey, compact stone, with one harder course, 0.3 m thick, in a pit [8300 4709] at Water Farm, Corsley, north of the district.

A section [8027 3993] near Dunkerton Cottages, Maiden Bradley, records the lowest part of the member; the junction with the Gault is sharp (Jukes-Browne and Scanes, 1901):

	Thickness m
CANN SAND	
Sand, yellowish, fine-grained, micaceous, passing down into rubbly and sandy Malmstone	2.44
Malmstone, buff-coloured, blocky, fossiliferous	3.05
Gault	
Clay, greenish, sandy, micaceous and glauconitic	0.61–0.91
Clay, dark grey to black	seen 3.05

Gault/Upper Greensand 'transition beds' at Fovant [SU 002 290] in the Salisbury district to the east, and the upper beds of the Gault at Fontmell Magna in the Shaftesbury district, yielded an *inflatum* Zone, *varicosum* Subzone fauna (Bristow and Owen, 1991; Mottram, 1957), establishing a maximum age for the base of the Cann Sand. Few fossils have been identified from the Cann Sand. Jukes-Browne and Scanes (1901) found '*Ammonites rostratus*' [= ?*Mortoniceras inflatum*] in 'Malmstone' [about 7960 4013] north of Kate's Bench Farm, Maiden Bradley. Silty sand and sandstone near the base of the Upper Greensand, possibly the Cann Sand, at Dilton Vale Farm north of the district (Mr G A Kellaway, manuscript, BGS) yielded *Mortoniceras (Pervinquieria)* sp. of the *pricei-inflatum* group, indicative of the *varicosum* Subzone. Evidence from the Shaftesbury Sandstone (see below) suggests that the upper part of the Cann Sand also falls in the *varicosum* Subzone.

Shaftesbury Sandstone (Member)

The Shaftesbury Sandstone (Bristow, 1989a) overlies the Cann Sand. The base is not exposed, but is marked by the sharp negative feature break below the escarpment. The member forms a prominent escarpment, capped by the Boyne Hollow Chert, along the Upper Greensand outcrop. A quarry [8737 2227] at Boyne Hollow, south of Shaftesbury, is designated the type section (Bristow et al., 1995).

North of the Mere Fault, the member is mostly between 15 and 18 m thick, but locally it is up to 25 m thick; between Upton and Milton [around 875 315], it is about 20 m thick, and forms well-featured ground; it is about 20 m thick near Donhead St Mary in the south-east.

The member consists of alternating beds dominantly of glauconitic coarse silt to fine-grained sand and weakly calcite-cemented sandstone. The sands vary from poorly to moderately sorted, with some beds well sorted. The member is capped by a hard, shelly, calcite-cemented, glauconitic sandstone. This uppermost bed has been variously referred to as the Ragstone Beds (White, 1923, p.46), Ragstone and Freestone Beds (White, 1923, p.51) and **Ragstone** (Drummond, 1970, fig. 2). The variable cementation in the beds below the Ragstone locally produces a nodular or rubbly sandstone. Where well developed, such as at Penselwood and Wolverton, the Ragstone was worked as building stone. A typical section in the upper beds was seen in a pit [8693 4129] at Longbridge Deverill (Jukes-Browne, manuscript; Woods and Bristow, 1995):

	Thickness m
Soil with weathered fragments of chert	0.91
BOYNE HOLLOW CHERT	
1. Cherty stone, white, weathered and broken	0.46
SHAFTESBURY SANDSTONE	
2. Sand, green with a layer of large calcareous concretions and some smaller ones	0.46
3. Sand, grey, very coarse, with rough, irregular calcareous masses full of shells (*Pycnodonte (Phygraea?) vesiculosum*, *Merklinia* cf. *aspera* and *Neithea gibbosa*)	0.61
4. Sand, green, sharp, less coarse, but with some large grains, indistinctly bedded	3.05

Fossils, particularly age-diagnostic ones, are rare in the Shaftesbury Sandstone; those listed in the section above may indicate the *auritus* Subzone. *Pycnodonte (Phygraea) vesiculosum* is particularly common in the Ragstone. Ammonites from the Shaftesbury district (Bristow et al., 1995) suggest an age no older than the *varicosum* Subzone for most of the member, although the Ragstone may be of *auritus* Subzone age, as suggested Drummond (1970), who regarded it as equivalent to the Potterne Rock (Owen, 1976).

Boyne Hollow Chert (Member)

The Boyne Hollow Chert (Bristow, 1989a) overlies the Shaftesbury Sandstone and caps the Upper Greensand escarpment. Despite extensive quarrying for road metal, there are few exposures. It was formerly well exposed at the type locality [8737 2227] at Boyne Hollow, Shaftesbury (Jukes-Browne and Hill, 1900), and in pits south of Warminster (see below and Woods and Bristow, 1995).

The thickness of the Boyne Hollow Chert varies from 8 to 20 m. East of East Knoyle, the member is 9 m thick (Jukes-Browne and Hill, 1900); at Wolverton, a well [7880 3164] proved a minimum thickness of 9.3 m.

The member consists of glauconitic quartz sand and sandstone with cherty and siliceous concretions and, in places, beds of chert up to 0.6 m thick. The basal bed, up to 1 m thick, of fossiliferous, glauconitic sand and weakly cemented sandstone with phosphatic nodules rests, possibly disconformably, on the Ragstone

A representative section was exposed in a pit [8193 3994] near Baycliff Farm (Jukes-Browne and Scanes, 1901; Woods and Bristow, 1995):

		Thickness m
Topsoil		0.46

BOYNE HOLLOW CHERT
A.	Silt, marly, pale buff, with angular fragments of chert	0.36
B.	Chert bed, pale grey	0.15
C.	Silt, marly, buff, sparsely glauconitic (*Neithea aequicostata*, *N. gibbosa*, *N. (N.) quinquecostata?*, *Holaster laevis* and *?Tiaromma michelini*)	0.61
D.	Sandstone, grey, spiculiferous	0.15
E.	Sand, glauconitic, greenish grey, soft, fine-grained, and laminated; small, irregular, spongiform concretions with many sponges	0.41
F.	Sandstone, grey, glauconitic, spiculiferous with irregular seam of yellowish sand and granular sandstone; many crabs	0.81
G.	Sandstone, calcareous, spiculiferous	0.38
H.	Sand, glauconitic, greenish grey, with large irregular masses of chert	0.91
I.	Silt, marly, greyish white	0.36

SHAFTESBURY SANDSTONE?
J.	Sandstone, grey, calcareous (*Merklinia* cf. *aspera*, *Neohibolites* sp., *Discoides subuculus*)	0.46

Bed A was equated with Bed 5 of Jukes-Browne and Hill's (1900) section at Maiden Bradley (see below).

For mapping purposes, the top of the member is taken above the highest chert bed or nodule, but up to 1 m of chert-free sandstone may intervene between the highest chert bed and the fossiliferous phosphatic nodule bed at the base of the Melbury Sandstone, as shown by Bed 4 in the following section [7980 3891] at Maiden Bradley (Jukes-Browne and Scanes, 1901):

		Thickness m

MELBURY SANDSTONE
1.	Marl, glauconitic ('Chloritic Marl'), with scattered phosphatic nodules and fossils and a sandy base	0.61
2.	Sand, brownish, glauconitic, with phosphatic concretions and fossils at base	0.15
3.	'Cornstones', phosphatised calcareous concretions with glauconitic sand matrix	0.30

BOYNE HOLLOW CHERT
4.	Sand, glauconitic, with some calcareous concretions	0.84
5.	Sand, greyish white, with siliceous sponge spicules and grey chert nodules	1.07
6.	Sand, glauconitic, grey, fine-grained, with large echinoderms and broken *Neithea*	0.61
7.	Chert, large blocks	0.46
8.	Sand, glauconitic, grey, fine-grained	0.30
9.	Sandstone, hard, granular, with siliceous sponge spicules	0.53
10.	Sand, pale grey	seen

From Bed 4, they recorded an Early Cenomanian, *mantelli* Zone fauna (see Woods and Bristow, 1995). This fauna came from the top few centimetres of the bed, with the matrix of Bed 3 passing 'down into the sand below'. A similar situation is described by Smart (1955) in the south of the Shaftesbury district, where Cenomanian fossils infill cracks and hollows in the phosphatised top of the Upper Greensand. The above fauna was the basis of Kennedy's (1970, fig.4) and Wright and Kennedy's (1984) dating of all of Bed 4, which they wrongly equated with the Rye Hill Sand, as Early Cenomanian. However, the record in Bed 4 of an indigenous Upper Albian ammonite, *Callihoplites vraconensis* (BGS specimen Zb 1213 determined by Dr H G Owen, but now lost), shows that some, if not most, of Bed 4 is of Albian age.

The chert nodules, up to 0.6 m thick, generally consist of a brown, black or grey, massive, flinty core (not always present) surrounded by up to 5 mm of more porous, white, yellow or brown siliceous material, set in a matrix of silt and fine-grained, calcareous, sparsely glauconitic sand. The nature and origin of the chert is discussed by Tresise (1961).

Jukes-Browne and Hill (1900) obtained a fauna, including the ammonite *Idiohamites* aff. *elegantulus*, from a 0.9 m-thick sand with phosphatic pebbles at the base of the Boyne Hollow Chert south of Shaftesbury (Bristow et al., 1995). Exposures on the north side of Shaftesbury (Bristow et al., 1995) revealed unfossiliferous sand with phosphatic nodules, on Shaftesbury Sandstone. A similar sequence was noted in the Bridport district, north of Beaminster [4839 0317] (Wilson et al., 1958). This phosphatic nodule bed probably represents an erosive event and subsequent transgression recognisable across southern and eastern England and thought to be of *dispar* Zone age (Morter and Wood, 1983). Support for the *dispar* Zone age is provided by *Merklinia aspera*, which appears to be unknown below the Ragstone (late *auritus* Subzone), by *Idiohamites elegantulus*, which ranges from the *dispar* Zone to the Cenomanian *mantelli* Zone, and by the probable presence in the chert beds of the Shaftesbury district of *Catopygus columbarius* and *Holaster geinitzi*, which also range from the *dispar* Zone into the Cenomanian (Bristow et al., 1995). At Maiden Bradley, the chert beds have yielded the Late Albian ammonite *Callihoplites vraconensis* (see above). Other fauna from elsewhere in the district includes *Neohibolites*, *Cardiaster fossarius*, *Discoides subuculus*, *Epiaster lorioli*, *Holaster laevis?*, *H. nodulosus* and *Tiaromma michelini*. This is not inconsistent with a *dispar* Zone age, the same age as the chert beds of the Bridport district (Spath, 1923–1943, p.423), although Carter and Hart (1977, p.105, fig. 45, 46) date the Chert Beds as Cenomanian in age on foraminiferal evidence.

Melbury Sandstone (Member)

The term Melbury Sandstone was used by Drummond (1970) for strata between the Boyne Hollow Chert and the base of the Chalk in the Melbury area south of Shaftesbury. In this district, the Melbury Sandstone, was known, in part, as the poorly defined 'Warminster Greensand'.

In the absence of good exposure, or borehole control, it is difficult to give a precise figure for the thickness of the Melbury Sandstone, but it probably varies from 1 to 4 m in the district.

The type section is the old Melbury quarry [8753 2015], where the base of the Melbury Sandstone is taken at the base of a pebbly, shelly, phosphatic bed (Bristow et al., 1995). In most sections in the district, the basal boundary is fairly sharp, but in the field it is not always possible to distinguish brash of the pebbly basal bed from the underlying glauconitic sandstone (locally up to 1 m thick) at the top of the Boyne Hollow Chert (Bed 4 at Maiden Bradley — see above).

The Melbury Sandstone consists of fine-grained, fossiliferous, calcareous sand and sandstone. In general, it comprises a lower unit of sand with common, very hard, irregular sandstone nodules — 'Cornstones' or 'Popple Bed', overlain, in the area east of Maiden Bradley, by similar sand without Cornstones and known locally as the **Rye Hill Sand.** At Rye Hill Farm (Figure 27), the following section was seen in two pits [8485 4017 and 8488 4017] in the lane south of the farm:

	Thickness m
MELBURY SANDSTONE	
Glauconitic Marl	
1. Marl, sandy and calcareous sand, with scattered brownish phosphates; hard and set like mortar, passing down into	0.91
Rye Hill Sand	
2. Sand, sharp greenish grey, slightly calcareous, many fossils (see Woods and Bristow, 1995)	0.76
Cornstones	
3. Sand, green, including scattered lumps of hard white calcareous stone	0.30

	Thickness m
BOYNE HOLLOW CHERT	
4. Sand, greenish grey, soft; scattered calcareous concretions	0.70

At some localities, for example the A303 cutting at Zeals [7856 3134] (Bristow et al., 1992) and a section [8103 3295] (Barton, 1994) near Mere, the basal conglomerate rests with a sharp break on the Boyne Hollow Chert. In other places, such as pits at Dead Maid (Bed 6, see below), Maiden Bradley (Bed 4, see above) and Rye Hill (Bed 4, see above), there is a sand beneath the Cornstones that was earlier included with the 'Warminster Greensand', but in this account, we regard this sand, of Albian age, as the top bed of the Boyne Hollow Chert (see above). The base of the Cornstones probably marks the Albian/Cenomanian stage boundary.

Overlying the Rye Hill Sand and/or Cornstones across the area is the poorly defined 'Chloritic' or 'Glauconitic' Marl, the identification of which presents special problems (Woods and Bristow, 1995) (see also Figure 27), since most sections are no longer extant and the lithological descriptions are not always reliable. For example, at Dead Maid Quarry [803 323], Bed 4 in the section below, referred to in the literature as a 'glauconitic marl', on re-examination, proves to be a glauconitic sandstone and clearly is part of the Melbury Sandstone (Barton, 1994). Bed numbers in the following account are from Jukes-Browne and Scanes (1901):

	Thickness m
WEST MELBURY MARLY CHALK	
1. Topsoil and yellowish chalky earth	1.37
2. Chalk marl, soft, passing down into hard, rocky, chalk marl	1.22
3. Marl, glauconitic, sandy, with a few phosphatic nodules	0.23
MELBURY SANDSTONE	
4. Marl, hard, glauconitic, sandy with phosphatised fossils and phosphatic nodules at base	0.61
5. Sandstone, brown and greenish brown, calcareous, with hard calcareous concretions ('Popple-bed') and phosphatised clasts and fossils	0.46
BOYNE HOLLOW CHERT	
6. Sandstone, hard, calcareous, glauconitic, with sparse concretions	0.46
7. Sand and sandstone, yellowish, sparsely glauconitic, with angular nodules and beds of brownish black chert	7.01

Bed 4 or its equivalent has been recognised during the resurvey at Zeals (Figure 27, bed B) [7856 3134] (Bristow et al., 1992) and near Mere [8103 3295] (Barton, 1994), and it suggests that the 'Chloritic' or 'Glauconitic Marl' at Search Farm [7930 3449], Norton Ferris [7940 3651], Maiden Bradley [7980 3891], Shute [8403 4054] and Rye Hill [8485 4017] should also be included as the topmost unit of the Melbury Sandstone (Woods and Bristow, 1995). This interpretation is also supported by the matrix of glauconitic sandstone on many specimens supposedly from the Chloritic/Glauconitic Marl. Accordingly, in this account, the term 'Glauconitic Marl' in quotation marks is used for most sandy occurrences between the top of the Rye Hill Sand or Cornstones and base of the West Melbury Marly Chalk, and included in the Melbury Sandstone. **Glauconitic Marl** is restricted to such units as Bed 3 at Dead Maid Quarry (see Woods and Bristow, 1995). The top of the Melbury Sandstone is taken where marl becomes a significant part of the sequence, but it is transitional in places. Almost certainly, there is a complex internal stratigraphy with pebble beds and phosphatic nodules marking erosional events.

The detailed biostratigraphy of the Melbury Sandstone is poorly known as it is difficult to distinguish derived from indigenous fossils. Some ammonites, such as *Anisoceras auberti* and *Scaphites (S.) equalis* have so far only been recorded in the Cornstones; others, such as *Forbesiceras largilliertianum*, *Hyphoplites* aff. *costosus*, *H. falcatus aurora*, *H. falcatus interpolatus*, *Mariella cenomanensis*, *Scaphites (S.) obliquus*, *Schloenbachia coupei* and *Sciponoceras roto* have only been found in the 'Glauconitic Marl'. The dominant components of the fauna indicate the *mantelli* Zone, both *carcitanense* and *saxbii* subzones, but there is also unequivocal evidence of the *dixoni* Zone low down in the Melbury

Sandstone in the Dead Maid pit [803 323]. *Mantelliceras saxbii*, characteristic of, but not confined to, the *saxbii* Subzone, is recorded in the Cornstones at both Dead Maid and Norton Ferris, and in the overlying beds [Beds ?3 and 4] at the former locality (Woods and Bristow, 1995). Most of the brachiopod and bivalve faunas from the Rye Hill Sand have calcite shell preservation, and ammonites are only weakly phosphatised. Thus, it is probable that the fossils are indigenous and the occurrence of *Mantelliceras mantelli* at Rye Hill Farm (Woods and Bristow, 1995) confirms the *mantelli* Zone. Wright and Kennedy (1984) recorded the *M. couloni* horizon in the *mantelli* Zone at Rye Hill Farm. Much of the Rye Hill Sand is probably of *carcitanense* Subzone age, but it ranges up into the *saxbii* Subzone and possibly the *dixoni* Zone. Besides the evidence from the Dead Maid Quarry, other indications of *dixoni* Zone faunas include *Inoceramus virgatus* in a fine-grained glauconitic sandstone in a well at Rye Hill (Woods and Bristow, 1995), in a sandstone just above a basal conglomeratic bed in the A303 cutting at Zeals (Bristow et al., 1992), in a ditch near Donhead St Andrew (Bristow, 1990b), and, with ?*Mantelliceras dixoni*, in a sandstone near Mere (Barton, 1994).

If the *saxbii* Subzone and *dixoni* Zone ages are confirmed for the Melbury Sandstone, this is younger than in the

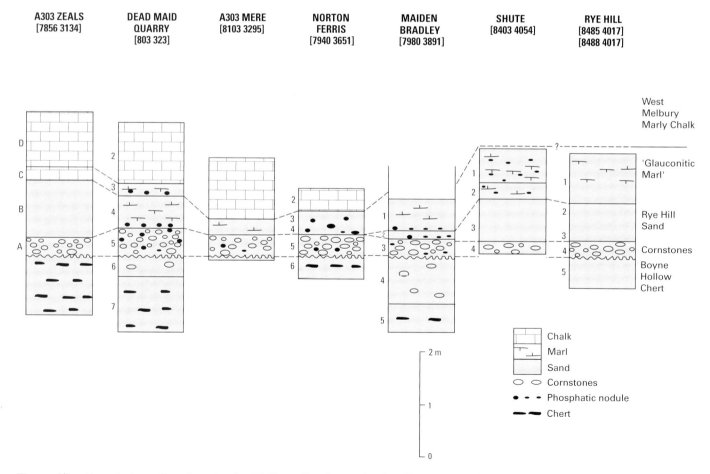

Figure 27 Correlation of sections in the Melbury Sandstone in the district. Numbers/letters adjacent to sections are locally applied field subdivisions (see also Woods and Bristow, 1995).

Shaftesbury area which is of *carcitanense* Subzone age (see also West Melbury Marly Chalk).

CHALK GROUP

The Chalk crops out over about 160 km^2 in the north-east of the district, with a small tract in the south-east. About 160 m of the group are preserved in the district. The Upper Chalk forms a prominent escarpment around the whole outcrop. In the west, outliers of Lower, Middle and Upper Chalk form conspicuous landmarks. The most prominent, Long Knoll (Plate 9) [786 376] to Little Knoll [810 378], south of Maiden Bradley, rises to 288 m OD, the highest point in the district. There is a general eastward dip of the Chalk, and a corresponding overall fall in ground level to just under 200 m OD in the east of the district.

The Wincanton district lay towards the northern margin of the Southern Province Chalk depositional basin (Mortimore, 1983) that covered southern England at that time; the northern margin was defined by the Mendip Axis and the London Platform.

The Chalk throughout much of southern Britain is divided into Lower, Middle and Upper formations, the base of the Middle and Upper Chalk is taken at the Melbourn Rock and Chalk Rock, respectively, both of which, are laterally persistent markers (see Bristow et al. (1997)).

By a combination of feature mapping, lithological variation and palaeontology, the Chalk of the Shaftesbury and Dorchester districts (Bristow et al., 1995) was divided into nine mappable units. Only the lower six units occur in the district (Table 12). In southern England, different names have been used for the Chalk successions in different places, but recently (Bristow et al., 1997), the Chalk of the whole of southern England has been reviewed and a unified nomenclature introduced (Table 12).

Lower Chalk (Formation)

The Lower Chalk principally crops out in an arc from Corton in the north-east to Mere, and from West Knoyle to Fonthill Bishop where the beds are more steeply dipping.

The junction with the Upper Greensand is fairly sharp and is marked by the appearance of marl and a rapid upward decrease in sand content.

The Lower Chalk is divided into the West Melbury Marly Chalk, overlain by the Zig Zag Chalk. The boundary between the two members is usually marked by an abrupt negative feature break, with the West Melbury Marly Chalk forming a low shelf, and the Zig Zag Chalk rising steeply from it (Plate 9). There is a marked difference in borehole gamma-ray signature between the two units (Bristow et al., 1995). The West Melbury Marly Chalk is equivalent to the lower part of the Chalk Marl of the South Downs succession; the Zig Zag Chalk is equivalent to the upper part of the Chalk Marl, together with the overlying Grey Chalk and Plenus Marls (Table 12, Figure 28).

WEST MELBURY MARLY CHALK (MEMBER)

The West Melbury Marly Chalk is up to 1 km wide at outcrop in the tract between Corton, Brixton Deverill and Mere. Because of steep dips, it has a very narrow outcrop between West Knoyle and Fonthill Bishop.

The thickness of the West Melbury Marly Chalk varies from 3 to 55 m, but lies mostly in the range 25 to 40 m. It is about 17 m thick in the area between Search Farm and Zeals Knoll north-west of Mere. A borehole [8570 3805] at Brixton Deverill, which commenced almost at the top of the member, proved 41.5 m. Where the member is thinnest (e.g. about 3 m thick west of Lower Barn, Horningsham [830 403]), it apparently results from channelling beneath the Zig Zag Chalk.

The member consists of soft, off-white, creamy and buff marly chalk, which is glauconitic and sandy in the basal part; there are a few thin harder beds of chalk. The base may be transitional with the Upper Greensand over a metre or so. The top of the member is taken at the top of the **Tenuis Limestone** (Bristow et al., 1997) which contains large *Inoceramus tenuis*, marks the first appearance in the succession, of *Acanthoceras*, and is associated with *Turrilites costatus*. In the district, it has been recognised only in a pit [828 402] west of Lower Barn Farm, Horningsham (Bristow, 1994), where it is 0.8 m thick.

Plate 9 Long Knoll from White Sheet Down. Long Knoll is an outlier of Zig Zag Chalk, Holywell Nodular Chalk and New Pit Chalk overlying low-lying ground formed by the West Melbury Marly Chalk and Boyne Hollow Chert that dip gently eastwards (to the right) in the foreground. (A15552)

Table 12 Chronostratigraphy and lithostratigraphy of the Upper Greensand and Chalk Group of the district.

Stages	Macrofossil		Southern England Chalk subdivisions former nomenclature		Wincanton district	
	Zones	Subzones				
SANTONIAN (pars)	Micraster coranguinum		Upper		Upper	Seaford Chalk
CONIACIAN	Micraster cortestudinarium		Upper	Top Rock	Upper	
	Sternotaxis plana			Chalk Rock		Lewes Nodular Chalk
TURONIAN	Terebratulina lata		Middle		Middle	New Pit Chalk
	Mytiloides labiatus s.l. (pars)					Holywell Nodular Chalk
	Neocardioceras juddii			Melbourn Rock		
CENOMANIAN	Metoicoceras geslinianum		Lower	Plenus Marls	Lower Chalk	(Plenus Marls)
	Calycoceras guerangeri			Grey Chalk		Zig Zag Chalk
	Acanthoceras jukesbrownei					
	Acanthoceras rhotomagense	Turrilites acutus	Chalk	Chalk Marl		West Melbury Marly Chalk
		Turrilites costatus				
	Mantelliceras dixoni					?
	Mantelliceras mantelli	Mantelliceras saxbii			Upper Greensand	Melbury Sandstone
		Neostlingoceras carcitanense		Glauconitic Marl		
				Warminster Greensand		
ALBIAN	Stoliczkaia dispar	Unnamed				
		Mortoniceras (Durnovarites) perinflatum		UPPER GREENSAND (pars)		Boyne Hollow Chert
		Mortoniceras (M.) rostratum				

Figure 28 Schematic stratigraphy of the Lower Chalk.

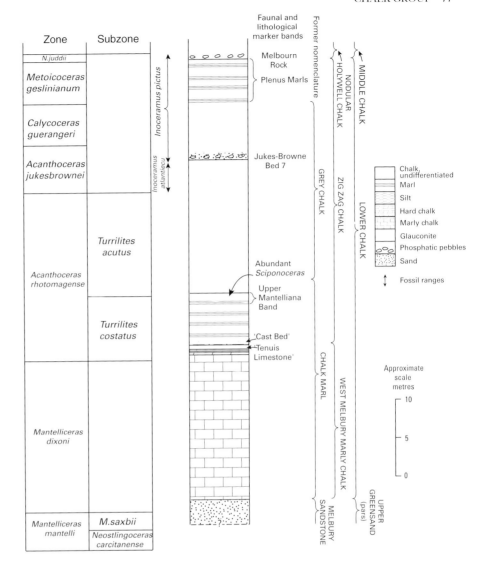

Apart from the basal beds, which were exposed at the A303 Zeals cutting [7856 3134], Dead Maid Quarry [803 323], A303 Mere pit [8103 3295], Search Farm [7930 3449] and Norton Ferris [7940 3651], there are few sections in the member (Barton, 1994; Freshney, 1995b). Fossils, commonly worn, fragmentary and phosphatised, include *Grasirhynchia grasiana*, *Inoceramus* ex gr. *crippsi* (including *I. crippsi hoppenstedtensis*), *I.* ex gr. *reachensis*, *I.* ex gr. *virgatus*, *Hypoplites curvatus*, *H.* ex gr. *falcatus*, *Hypoturrilites tuberculatus*, *Mantelliceras* sp., *Mariella (M.) lewesiensis*, *Parapuzosia (Austiniceras)* sp., *Puzosia* sp., *Schloenbachia varians*, *Turrilites costatus*, *T.* aff. *wiestii?* and *Catopygus columbarius*. Collectively, the fauna indicates the *mantelli*, *dixoni* and early *rhotomagense* zones, but it is most probable that the *mantelli* Zone faunas are derived (Woods and Bristow, 1995). Common and large *Inoceramus* ex gr. *virgatus* at the A303 Mere section [8103 3295] (Barton, 1994; Woods, 1994c, unpublished report*) compare with the acme occurrence of this bivalve at Folkestone, in the lower part of the *dixoni* Zone (Gale, 1989), and with those from the **Dixoni Limestone** of the Chilterns (Wood, 1990, unpublished report*).

ZIG ZAG CHALK (MEMBER)

The Zig Zag Chalk comprises 10 to 30 m of firm white chalk. It is 20 to 25 m thick on the western side of White Sheet Downs and along Long Knoll. The member has a narrow outcrop at the foot of the Chalk escarpment. The base is taken at the top of the Tenuis Limestone and forms a marked negative feature break at outcrop.

No section in the Wincanton district exposes the whole of the Zig Zag Chalk, although the pit [9350 2395] at the foot of White Sheet Hill in the south-east (Bristow, 1990b; Jukes-Browne and Hill, 1903) must formerly have exposed a good section. However, there are several sections in the Shaftesbury district to establish that many elements of the stratigraphy established in Sussex and Kent can be recognised in that district (Bristow et al., 1995) (Figure 28). Research on Middle Cenomanian sections in Sussex and Kent (Gale, 1989; 1990) revealed a consistent stratigraphy in the lower part of the *rhotomagense* Zone. The Tenuis Limestone is overlain by the **'Cast Bed'**, a bed of brown, silty chalk with a rich fauna of poorly preserved, aragonite-shelled molluscs, notably gastropods, common

Oxytoma seminudum, and an assemblage of small brachiopods including *Modestella geinitzi*. The 'Cast Bed' broadly equates with the **Totternhoe Stone**; it is overlain by alternating marls and marly chalk with common *Orbirhynchia mantelliana*, comprising the highest (**Upper Mantelliana Band**) of the three *O. mantelliana* bands of southern England. The sequence is capped by a limestone with abundant *Sciponoceras baculoides*, possibly seen at Search Farm [7930 3449] (Woods and Bristow, 1995), the top of which marks the mid-Cenomanian nonsequence (Carter and Hart, 1977) and the base of the *acutus* Subzone. A rhythmic alternation of marls and marly limestones continues for 1 to 2 m above the Upper Mantelliana Band, above which there is a change to more massively bedded chalk with thin marl bands. Possible occurrences of the Upper Mantelliana Band were in pits near Search [7930 3449] and Lower Barn farms [8280 4021 and 8282 4022].

The upper part (Grey Chalk equivalent) of the Zig Zag Chalk consists of firm white chalk, divisible into a lower unit, up to **'Jukes-Browne Bed 7'** (see Figure 28), with common *'Inoceramus' atlanticus* in the upper part, and an upper unit with *I. pictus*. *'T.' atlanticus* has a restricted range from the base of the *jukesbrownei* Zone to the base of Jukes-Browne Bed 7; *I. pictus* s.l. ranges from within the *jukesbrownei* Zone to the top of the Plenus Marls, with related species extending into the Holywell Nodular Chalk (Figure 28). A representative section of the Zig Zag Chalk, including Jukes-Browne Bed 7, in the pit at the western end of White Sheet Hill [9352 2398] shows:

	Thickness m
Chalk, hard, rubbly weathering, slightly gritty with pycnodonteine oyster	1.00
Chalk, off-white, nodular and gritty; *Acanthoceras jukesbrownei*, together with '*Inoceramus*' *atlanticus*, occurs 0.65 and 1.45 m above the base [=Jukes-Browne Bed 7]	2.50
Marl, pale grey, silty with '*T.' atlanticus*; sharp junction [equivalent to marl marking lower limit of Jukes-Browne Bed 7]	0.15
Chalk, off-white, massive, blocky weathering with conchoidal fracture with common '*T.' atlanticus* for at least 1.5 m below the marl	2.00

Elsewhere, the higher part of the Zig Zag Chalk is only poorly exposed. *I. atlanticus* has also been found at Search Farm [7913 3459] and on White Sheet Downs [8099 3340, 8096 3341, 8097 3342]. *I. pictus* was found south-west of Sutton Veny [9005 4118]. Jukes-Browne Bed 7 was also seen in the track [8095 3342] up White Sheet Hill, Mere, and near Corton, where glauconitised pebbles were noted [9327 4134 and 9307 4126] about 3 m above the base of the Zig Zag Chalk.

The ammonite *Acanthoceras*, including *A. jukesbrownei* and *A. rhotomagense*, is fairly common and has been found on White Sheet Downs [8099 3340], near Search Farm [7913 3459], close to the base of the member [8906 4125] and 5 m higher [8985 4131] on the north-east side of Cow Down, at Lower Barn Farm [827 402], and on White Sheet Hill [9350 2395]. *Turrilites costatus* has been found in brash [8182 3906] on the lower slopes of Brimsdown Hill and at Lower Barn Farm [8279 4021].

The **Plenus Marls** are included in the top of the Zig Zag Chalk, because it is not possible to map them separately. Although rarely well exposed, they are present along the whole outcrop of the district and probably have a similar lithology to those of the Shaftesbury district. There, the Plenus Marls consists of an alternating sequence of blocky, white chalk in beds up to 1.2 m thick, and beds of medium grey silty marl, mostly between 1 and 10 cm thick, though the highest (Jefferies' Bed 8) can be up to 50 cm thick (Bristow et al., 1995). The Plenus Marls have a distinctive gamma-ray signal and are readily recognisable in geophysical logs. During this survey, greenish grey marl and marly chalk characteristic of the Plenus Marls was seen south-west of Tytherington [9116 4069 and 9114 4062], on Tytherington Hill [9186 4022], at West Knoyle [8664 3268], Littlecombe Bottom [9092 3986], and near Fonthill House, where the marl yielded *Actinocamax plenus* [9465 3255].

In a study of the Plenus Marls in the Anglo-Paris Basin, Jefferies (1963) divided them into eight beds, each with a distinctive fauna, which he recognised throughout southern England. At the base of each marl seam, there is an erosion surface, the most strongly developed is the 'sub-*plenus*' erosion surface at the base. The Plenus Marls, coextensive with the greater part of the *geslinianum* Zone, contain common *Inoceramus pictus*.

Middle Chalk (Formation)

The base of the Middle Chalk is taken at the base of the **Melbourn Rock**, a feature-forming unit that consists of up to 3 m of hard, nodular, poorly fossiliferous chalk, overlain by shell-detrital and shelly chalk. The Melbourn Rock forms a marked lithological contrast to the underlying Plenus Marls, and is one of the easiest of the Chalk boundaries to map and to recognise on geophysical logs.

The Middle Chalk is divided into two members, Holywell Nodular Chalk overlain by New Pit Chalk. The boundary between the members is fairly sharp in sections and boreholes, but is difficult to map precisely in the field.

In this district, the Middle Chalk maintains a constant thickness of about 25 m. A well [8830 3652] at Lower Pertwood Farm proved 24.2 m of Middle Chalk (Jukes-Browne and Hill, 1903).

HOLYWELL NODULAR CHALK (MEMBER)

The Holywell Nodular Chalk consists of 10 to 20 m of mainly nodular chalk, with some weak chalkstones, thin marl seams and, in the higher parts, smooth-textured chalk. It is characteristically rich in *Mytiloides* species, either fragmented or entire, and which are common everywhere in the brash.

The member forms a readily identifiable and mappable unit, and has a narrow outcrop usually in the face of the main scarp. In places, mainly on the less steep slopes, it forms a positive feature. Locally, such as north-east of Mere and south-west of Tytherington, the nodular chalks form a subsidiary escarpment below the main scarp.

The inoceramids of the member show a succession of species. At the base, and continuing up from the Zig Zag Chalk to the top of the Cenomanian part of the Holywell Nodular Chalk, is *Inoceramus* ex gr. *pictus*, such as found north-east of Brixton Deverill church [8672 3909]. The Turonian part is characterised by *Mytiloides* species, including *M.* cf. *labiatus*, *M. mytiloides* and *M. submytiloides*.

The member is 15 m thick at the western end of White Sheet Downs and along Long Knoll between Mere and Maiden Bradley.

There is no good section through the Holywell Nodular Chalk, but pits at East Knoyle [8875 3109] (Jukes-Browne and Hill, 1903) and White Sheet Hill [9357 2394 and 9359 2401] once exposed the whole sequence.

NEW PIT CHALK (MEMBER)

The New Pit Chalk, 10 to 16 m thick, comprises smooth, firm, white chalks with marl seams; it is 15.2 m thick at East Knoyle [8875 3109] (Jukes-Browne and Hill, 1904, p.75). The base is taken where firm, smooth chalk appears in the succession. In Sussex, this is coincident with the Gun Gardens Main Marl (Mortimore, 1986). Following

Bristow et al. (1997), the upper part of the newly defined New Pit Chalk includes the Glynde Marls, which were formerly (Mortimore, 1986) taken as the base of the Lewes Nodular Chalk. *Mytiloides* species (of the *hercynicus–subhercynicus* group) are typically thin-shelled, or with no preserved shell, and of a 'flattened' morphology. Brash of firm, slabby chalk with the above species was noted north-north-east of Kingston Deverill church [8482 3755], south of Fir Clump [8560 3878; 8570 3863], on the east side of Tytherington Hill [9203 3996], east of Knook [9437 4188], on the south-east side of Corton Hill, on White Sheet Hill [9379 2402] and in the road bank [8840 3116] north of East Knoyle.

Groups of well-developed marl seams, notably the New Pit Marls near the top, characterise the New Pit Chalk in Sussex, and give distinctive geophysical signatures. In this district, there are few good sections; marl seams were noted in the Charnage Down pit [8365 3285]. Grey flints occur in the upper part of the member, such as exposed in tracks up Corton [9434 3975] and White Sheet [9379 2402] hills.

In addition to *Mytiloides* species, *Collignoniceras woollgari* and *Lewesiceras peramplum* were found on White Sheet Hill [9379 2402]. In the Charnage Down Pit [8365 3285], fauna from the top of the member includes *Inoceramus cuvieri*, consistent with a level in the *lata* Zone. The New Pit Chalk spans the *mytiloides* spp./*lata* zonal boundary.

Upper Chalk (Formation)

The base of the Upper Chalk was originally defined by the base of the Chalk Rock, but as redefined by Bristow et al. (1997), it is taken at the base of hard, nodular chalk, the Lewes Nodular Chalk, above smooth chalks of the New Pit Chalk. The base coincides with the base of the Chalk Rock Formation of Bromley and Gale (1982) and the base of the Spurious Chalk Rock in Dorset and the Isle of Wight. Flints are common in much of the Upper Chalk. The base of the Upper Chalk can be recognised with ease both in the field and in boreholes across southern England; it has a characteristic signature on sonic logs (Bristow et al., 1997, fig. 9), although the style of signature varies according to the depositional setting (Mortimore and Pomerol, 1987).

By a combination of lithology and feature mapping, the Upper Chalk can be divided into up to six mappable members, of which only the lowest two occur in the Wincanton district (Table 12).

Lewes Nodular Chalk (Member)

The Lewes Nodular Chalk equates with the combined Akers Steps and St Margarets members of Robinson (1986) in the North Downs succession. Its base forms the revised base of the Upper Chalk. The member usually forms a good positive feature break at the top of the escarpment; in places, for example, White Sheet Hill, it forms extensive hill caps. The onset of nodularity coincides approximately with the appearance of common flints throughout the succession in southern England.

The basal Upper Chalk becomes condensed towards the northern margin of the Southern Province deposi-

tional basin (Mortimer, 1983) and the nodular chalks are replaced progressively by chalkstones and superimposed mineralised (glauconitised and/or phosphatised) hardgrounds comprising the **Chalk Rock**. This change from nodular chalk and hardgrounds to chalkstone with glauconitised hardground surfaces occurs across the northern part of the Shaftesbury district (Bristow et al., 1995). In the south of the Wincanton district, on White Sheet Hill [9375 2380], the Chalk Rock, with four glauconitised hardgrounds is 5 m thick (Bromley and Gale, 1982). Northwards, towards the Mere Fault, the Chalk Rock thins to about 2 m near Hindon [9178 3217]. North of the Mere Fault, from Mere [8367 3292] to the northern margin of the district, it is just over 1 m thick. The Chalk Rock is excellently displayed in the Charnage Down Quarry (Figure 29), where about 1.7 m of nodular chalk, the basal unit of the Lewes Nodular Chalk, underlies porcellanous and glauconitised chalkstones of the Chalk Rock (Plate 10).

The base of the Chalk Rock is diachronous, ranging from the lowest (Ogbourne) hardground in the type area at Ogbourne Maizey on Marlborough Downs and Charnage Down in this district, to the highest (Hitch Wood) hardground in the north Chilterns (Bromley and Gale, 1982).

Plate 10 Chalk Rock at the top of the section overlying nodular chalk at the base of the Lewes Nodular Chalk, Charnage Down Pit. Hammer length 30 cm. (GS482)

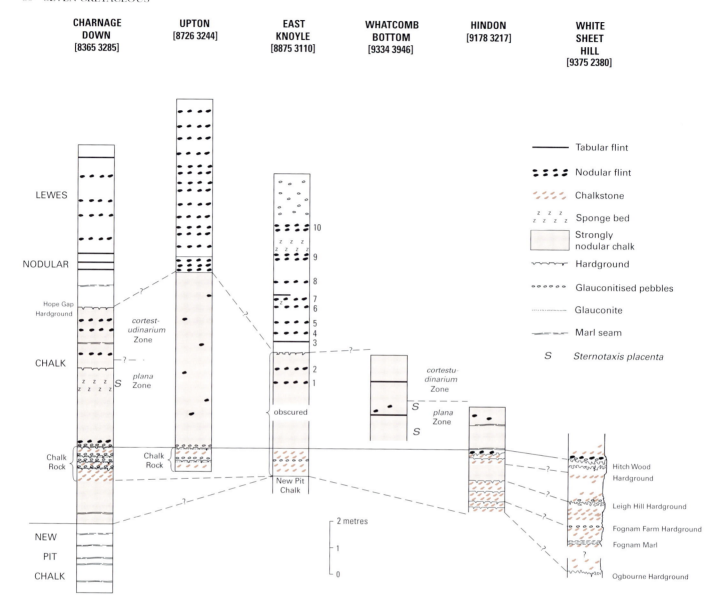

Figure 29   Correlation of Lewes Nodular Chalk sections.

The **Hope Gap Hardground** occurs 5 to 6 m above the top of the Chalk Rock (Figure 29); it is another well-developed hardground, but not so strongly mineralised. The fauna from the Charnage Down and East Knoyle [8875 3110] pits (Woods, 1992, unpublished report*; 1994c, unpublished report*) establishes a correlation with the type area in Sussex (Mortimore, 1986) and to part, at least, of the **Top Rock** in the Chilterns–East Anglia area.

The Lewes Nodular Chalk comprises the higher part of the *lata* Zone, the overlying *plana* Zone (Chalk Rock to Top Rock), and a variable amount of the *cortestudinarium* Zone (Table 12). The top of the member is slightly diachronous; in the Shaftesbury district, it may fall locally in the basal *coranguinum* Zone (Bristow et al., 1995), but in the Wincanton district it lies in the mid-*cortestudinarium* Zone (see Seaford Chalk below).

SEAFORD CHALK (MEMBER)

The term Seaford Chalk was introduced in Sussex by Mortimore (1986) for firm, white, flinty chalk with few marl seams (except near the base), overlain by similar chalk with common marl seams, the Newhaven Chalk. Only the Seaford Chalk crops out in the Wincanton district. The member consists of least 60 m of firm, white chalk with regular courses of flint nodules.

The base is diachronous across Wiltshire and Dorset; in this district, it falls in the middle of the *cortestudinarium* Zone, in contrast with the Shaftesbury area (Bristow et al., 1995) and Sussex, where it lies at the base of the *coranguinum* Zone. Only the Coniacian part of the *coranguinum* Zone is present in the district.

A borehole [9353 3448] at Chicklade Bottom proved 16.3 m of soft to firm white (Seaford) chalk of *cortestudi-*

narium Zone age at a level above the Hope Gap Hardground (Bristow, 1995c). Brash near the borehole yields common *Platyceramus* and *Volviceramus* (including *V. involutus*). In Sussex, *V. involutus* appears about 6 m up in strata of *coranguinum* Zone age, and ranges up for a further 15 m before becoming extinct; it is particularly common at the level of the Seven Sisters Flint (Mortimore, 1986, fig. 16). In the Chicklade and Brixton Deverill areas, several *Volviceramus* are preserved in flint [9026 3415, 9097 3497, 9487 3533, 8664 3791], possibly from one level, and may, therefore, be the Seven Sisters Flint. In the present district, most occurrences of *Volviceramus* fall within the range of 20 to 50 m above the base of the Seaford Chalk (ie. 2 to 32 m above the base of the *coranguinum* Zone). Abundant fragments of the inoceramid *Platyceramus* occur in brash all over the district and are characteristic of much of the Seaford Chalk.

Details of the many sections and occurrences of brash can be found in the Open-file reports listed in Information Sources (p.92). A summary of the more important localities is given below.

A303 [7853 3130], Zeals. Basal beds and junction with Upper Greensand formerly exposed (Bristow et al., 1992; Woods, 1991, unpublished report*).

Dead Maid Quarry [8035 3230]. Strata, of *mantelli* and *dixoni* Zone age, exposed at the top, above Melbury Sandstone and Boyne Hollow Chert (Bartlett and Scanes, 1917, fig. 10; Woods, 1994c, unpublished report*; 1995, unpublished report*).

Pit [8104 3291] alongside the A303, Mere. Beds of *dixoni* Zone age exposed (Barton, 1994; Woods, 1994c, unpublished report*).

Pits [7949 3407 and 7889 3466] at Search Farm; basal beds exposed (Freshney, 1995b; Woods and Bristow, 1995).

Pit [7940 3651] at Norton Ferris; beds of *dixoni* Zone age (Freshney, 1995b; Woods and Bristow, 1995).

Pit [827 402] at Lower Barn Farm straddles the Zig Zag/West Melbury Marly Chalk boundary; possible occurrence of Tenuis Limestone (Bristow, 1994; Woods, 1994c, unpublished report*).

Pit [9350 2395] at western end of White Sheet Hill; most of the Zig Zag Chalk, including Jukes-Browne Bed 7, and basal beds of the Holywell Nodular Chalk formerly exposed (Bristow, 1990b; Jukes-Browne and Hill, 1903).

Quarry [8875 3110] east of East Knoyle (Jukes-Browne and Hill, 1904 and Ms) once exposed a sequence from the basal Holywell Nodular Chalk to the Lewes Nodular Chalk; only the upper 9 m of the latter now seen (Bristow, 1995a, fig. 4; Woods, 1994c, unpublished report*) (Figure 29).

Charnage Down Quarry [8367 3292]: excellent sections from the top New Pit Chalk to above the 'Top Rock' (Bartlett and Scanes, 1917; Barton, 1994; Mortimore, 1987; Woods, 1992, unpublished report*; 1994c, unpublished report*) (Figure 29).

Mere Down [8279 3363, 8227 3378 to 8227 3370]: good sections in Chalk Rock with prominent, irregular and deeply burrowed, mineralised hardgrounds (Barton, 1994).

Track [8080 3421 to 8080 3417] up White Sheet Hill. Several hardgrounds, including Hitch Wood Hardground, exposed (Barton, 1994; Bromley and Gale, 1982, fig. 12; Woods, 1994c, unpublished report*).

New Pit Chalk, Chalk Rock and higher part of the Lewes Nodular Chalk on White Sheet Hill [8005 3511] (Bartlett and Scanes, 1917; Barton, 1994).

Chalk Rock [8181 3643] next to track up Rodmead Hill (Barton, 1994).

Trench on Brimsdown Hill [853 391] in Chalk Rock exposed 1 m of chalkstone with the Ogbourne, Fognam Farm and Hitch Wood hardgrounds (Bromley and Gale, 1982, fig. 12).

South-east of Monkton Deverill [8695 3660]: three glauconitised chalkstones (Bristow, 1995a, fig. 3).

Lord's Hill Farm [8805 3946]: porcellanous chalkstone, with glauconitised hardgrounds (Bristow, 1995a, fig. 3).

West side of Tytherington Hill [9118 3973]: nodular chalkstone, capped by a glauconitised hardground (Bristow, 1995c).

Lane [9278 3838] west side of Corton Down: chalkstone with glauconitised hardground; large and small forms of *Micraster leskei?* (Bristow, 1995c).

South side of Whatcomb Bottom [9334 3946]: 3.15 m section in high *plana* Zone and basal *cortestudinarium* Zone chalk below the level of the Hope Gap Hardground (Bristow, 1995c, fig. 3; Woods, 1994c, unpublished report*).

Quarry [8726 3244] in *plana* Zone chalk near Chapel Farm, Upton (Jukes–Browne, Ms) (Bristow, 1995a) (Figure 29).

Borehole [9353 3448] near Chicklade Bottom; 16.3 m of Seaford Chalk of *cortestudinarium* Zone age (Woods, 1993a, unpublished report*; Wilkinson, 1995d, unpublished report*;), on Lewes Nodular Chalk (Bristow, 1995c).

Road cutting [9092 3147] north-north-east of Ruddlemoor Farm: basal Lewes Nodular Chalk (Bristow, 1995c; Mottram, 1961, fig. 2).

Section [9178 3217] north side of the main road south-east of Hindon: Chalk Rock and associated strata (Bristow, 1995c, fig. 4; Woods, 1994c, unpublished report*) (Figure 29).

Pit [9375 2380] on top of White Sheet Hill (Bromley and Gale, 1982, fig. 10). Most of the pit filled; 1 m of the uppermost beds, including the Hitch Wood Hardground, remains (Bristow, 1990b) (Figure 29).

Section [8977 3847], Parsonage Down. About 2.5 m of firm, white, flinty chalk, with common *Platyceramus*, *V. involutus*, and *Micraster coranguinum?*, about 35 m above the base of the Seaford Chalk (Bristow, 1995a).

Pit [8619 3283] north-east of Manor Farm, West Knoyle; about 2 m of firm white chalk, with common *Platyceramus?* and *Volviceramus involutus*, at least 20 to 25 m above the base of the Seaford Chalk (Bristow, 1995a).

Borehole [9151 3517] north-east of Chicklade; 29 m of Seaford Chalk (Bristow, 1995c; Woods, 1993a, unpublished report*).

EIGHT

Structure

The Wincanton district lies at the north-western end of the Wessex Basin. The basin comprises a complex of structural highs and sub-basins. In this district (Figure 30), the Mendip High, and its southerly extension the Bruton–Norton Ferris High, is separated from the Mere Basin in the south-east by the Mere Fault. The Mere Basin is bounded on its southern side by the Cranborne–Fordingbridge High. This latter high extends from the southern margin of the district and across most of the Shaftesbury district as far as the Coker and Cranborne faults which form the northern boundary of the Winterborne Kingston Trough (Bristow et al., 1995, figs. 4 and 6).

The structural history of the district can be simplified into three broad periods, each characterised by a dominant tectonic regime.

Pre-Permian Devonian and Carboniferous strata have been proved in the Norton Ferris, Bruton and Fifehead Magdalen boreholes, and crop out about 2 km north and north-west of the district. In the Mendip Hills, these sequences are extensively deformed by large-scale folds and thrusts (Green and Welch, 1965; Williams and Chapman, 1986). Similar structures inferred from seismic data in this district, south of the Mere Fault (Chadwick et al., 1983), are consistent with a late- or post-Carboniferous compressional or transpressional tectonic regime (see Chapter 3).

Permian to Palaeogene Initial rifting of the Wessex Basin took place during the early Permian, with the formation of the Mere Basin and Winterborne Kingston Trough. The extensional tectonic regime was controlled mainly by east–west-trending faults, downthrowing to the south; stratal thickness changes markedly across these faults which are linked by a set of dextral north-north-west-trending strike-slip faults (Bristow et al., 1995). Both fault sets are probably controlled by reactivation of Variscan structures.

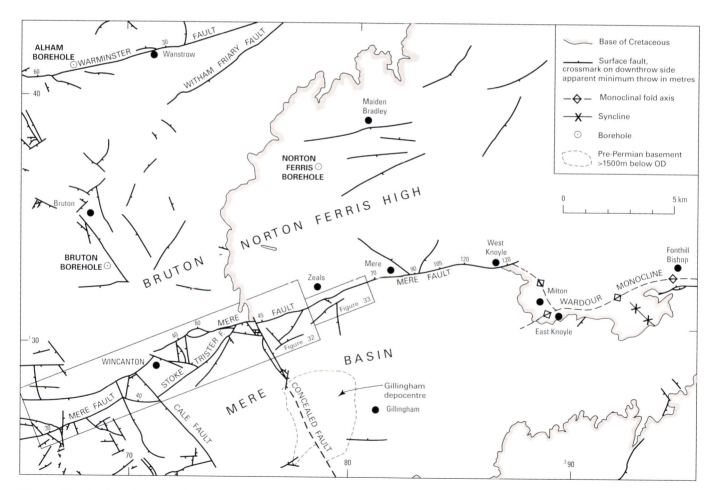

Figure 30   Surface faults and folds in the district and the location of Figures 32 and 33.

Both the Mere Basin and Winterborne Kingston Trough were active depositional centres during Permo-Triassic times. The Mere Basin was still in existence during deposition of the Lias and was affected by penecontemporaneous faulting. In contrast, during the Middle Jurassic and lower part of the Upper Jurassic, the Mere Basin was no longer a separate depocentre (Bristow et al., 1995, fig. 11) and little faulting or movement along existing faults took place. A period of extension during the Late Jurassic, well seen in the Wealden Basin to the east, has little expression in this district, except during the Late Oxfordian. At that time, deposition of the Corallian Group appears to have been controlled by syndepositional faults. Erosion occurred during the Early Cretaceous, with the progressive removal of some late Jurassic strata on either side of the fault. Erosion ceased with the Aptian/Albian transgression. This led to the burial of the major structures, such as the Mere Fault, by uninterrupted deposition of the Gault, Upper Greensand and Chalk.

Neogene   Compression or uplift of the Wessex Basin took place in a compressional tectonic regime which probably culminated in the Miocene. Steep, normal, east–west faults such as the Mere Fault were reactivated as high-angle reverse faults, while strike-slip faults record reciprocal directions of displacement.

The surface fault and fold distribution map (Figure 30) shows the major east-north-east-trending faults of the district.

WARMINSTER FAULT

The Warminster Fault [655 407 to 880 450] is at least 25 km long, extending from Maes Down [640 408], about 2 km west of the district, to Warminster which lies just to the north in the Frome district. It is near-vertical for much of its length, strikes generally east-north-east and down-throws to the north. West of the district, it becomes a steeply south-dipping reverse fault with a sinuous east–west trace. Displacement estimates indicate a maximum northerly downthrow of nearly 70 m at Small Down [665 408], decreasing westward to about 50 m in the Glastonbury district and eastward to 30 m in the Frome district where the fault cuts Lower Chalk.

The Lias succession thickens markedly southwards across the Warminster Fault, consistent with the overall southward thickening of this unit away from the Mendip Axis. The fault has a complex history which is illustrated by a comparison of thicknesses recorded in the Alham Borehole [6793 4118], just north of the fault, with those of the mapped succession to the south. In the borehole, the lower part of the Lower Lias, Pylle Clay and Spargrove Limestone, is represented by 7.6 m of massive limestones, whereas to the south, the same strata are about 17 m thick (Westhead, 1994). In contrast, in the borehole, the overlying sequence, the Ditcheat Clay (47 m), Pennard Sands (c.13 m) and Down Cliff Clay (14 m) is twice as thick, 74 m, compared with the outcrop to the south. This is due to anomalously thick Ditcheat Clay and Down Cliff Clay, even though the Pennard Sands reverses this pattern

and is up to 25 m thick south of the fault. The topmost member of the Upper Lias, the Bridport Sands, is absent in the borehole and occurs only as lenses up to 3 m thick north of the fault. It is up to 20 m thick south of the fault.

These thickness variations indicate that the Warminster Fault was active throughout the Early Jurassic, mostly as a south-throwing extensional structure, but during middle Pliensbachian to early Toarcian times, down-throw was to the north. Changes in thickness probably result from penecontemporaneous variations in deposition and erosion. In contrast, the Inferior Oolite and Great Oolite Group show little thickness variation across the Warminster Fault, indicating an absence of either structural relief or fault activity during the Aalenian to Bathonian.

The Warminster Fault was probably periodically reactivated in Late Jurassic and Early Cretaceous times (Callovian to Albian) with downthrow largely to the south, although this cannot always be determined due to the extensive pre-Cretaceous erosion that removed Upper Jurassic strata north of the fault. The Lower Cretaceous Cann Sand is thinner north of the fault than it is to the south.

The early movement across the Warminster Fault was largely down-to-the-south, whereas the total downthrow at crop is to the north. This indicates that there was a reversal of the fault during Neogene basin inversion. The displacement history is similar to that of the Mere Fault (see below).

WITHAM FRIARY FAULT

The Witham Friary Fault [717 395 to 768 430] has a down-throw of between 25 and 35 m to the south, creating a horst between it and the Warminster Fault. The down-throw decreases from a maximum of 35 m in the Kellaways Formation at Witham Friary; it dies out to the south-west in Forest Marble, near Upton Noble, and to the north-east in the Peterborough Member at Trudoxhill, north of the district, where it probably converges with the Warminster Fault.

Elsewhere, faults with a similar north-east trend control the orientation of part of the Brue valley. The northernmost, in the Forest Marble, is exposed [7104 3704] in river cliffs west of South Brewham, where strata are inclined 20°NW on the north-west side of the fault and about 50°NW on the south-east side, with an estimated downthrow of 5 to 10 m to the north-west. An exposure nearby [7088 3689] reveals a south-east-verging fold with an axis plunging at 07/040°, an axial surface inclined 060/44°NW and a steep southern limb, probably related to movement on the same fault.

MAIDEN BRADLEY FAULTS

Small faults south of Maiden Bradley may explain the large re-entrant and Long Knoll outlier of Chalk in that area. The faults trend east-north-east, broadly parallel to the Mere and Warminster faults. They offset the base of the Chalk, but do not cut the Middle and Upper Chalk.

Although throws at outcrop are less than 10 m, seismic reflection data show that they are much larger at depth.

The southernmost fault extends south-west in the subsurface, from just west of Kingston Deverill to Stourton [775 340], where it is cut by a north-west-trending fault along Six Wells Bottom. The latter, concealed, fault, with little or no displacement in the Upper Greensand, has a throw of almost 50 m north-east at the top of the Inferior Oolite. Similar concealed north-west-trending faults control the Gault re-entrants and inlier of Corallian strata between Stourton and Penselwood.

MERE FAULT

The dominant structural feature of the district is the Mere Fault (Edmunds, 1938, p.174) (the 'Great' Fault of Bartlett and Scanes, 1917, p.120), and associated Wardour Monocline in the east. The evolution of the Mere Fault and Wardour Monocline and a full interpretation of the movement history are detailed by Barton et al. (1998); only some of the more salient points are included in this account.

The main part of the fault extends east-north-east for more than 20 km across the district as far as West Knoyle. In the west, it forms a braided zone of fractures and *en échelon* faults, 1 to 2 km wide, with variable directions of downthrow (Figure 32). At East Knoyle, a buried north-east-trending fault is parallel with, and stepping-over about 3 km south from the Mere Fault.

Displacements along the length of the Mere Fault, derived from interpretation of seismic reflection data and from outcrop are plotted for the subsurface at Top Basement, Top Penarth Group and Top Inferior Oolite levels (Figure 31). Apparent stratal offsets change markedly along the length of the fault and vary according to stratigraphical or structural levels. The largest dip-slip displacements are in the range of 200 to 350 m at the Top Basement reflector (base of the Permo-Triassic), and occur only in a 12 km-long fault portion between Wincanton and Mere. This portion coincides with the thickest sedimentary strata in the Gillingham depocentre just to the south (Figure 30). Throws along the same segment at the Top Penarth Group reflector are in the range 100 to 200 m, although this reflector shows a local reverse displacement around Mere. Both displacement curves diminish to the west and east, suggesting that syndepositional movement in those areas was unimportant.

The displacement curve for the Top Inferior Oolite reflector shows smaller throws and changes along strike from down-to-the-south, nor-

mal displacement in the west, to reverse displacement with down-to-the-north throw, east of Leigh Common.

The Mere Fault at surface is described in two principal segments from west to east.

Wincanton segment

This segment is a complex zone in which numerous faults have relatively small normal or reverse throws (Figure 32). The same stratigraphical units are present across the fault zone and the magnitude of small downthrows can be estimated reliably. A double fault strand is present between Blackford and Holton. Seismic data show that it is the southern fault that displaces the Top Basement and Top Penarth Group reflectors, and it is this strand which is correlated with the Mere Fault. Downthrows along the southernmost strand range from 15 m to the south, to 30 m to the north, consistent with some post-Cretaceous displacement. The northern strand shows down-to-the-south throws that increase eastwards to 40 m near Wincanton. A similar increase in throw is also seen along the Stoke Trister Fault from less than 10 m south of Wincanton, to about 100 m near Stoke Trister.

Eastwards, between Wincanton and Bourton, the fault has a sinuous trace with downthrows between 40 to 60 m to the north. Farther east, in a short segment between the intersections with the Gillingham and Stoke Trister faults, the Mere Fault has 50 m of normal displacement. Between Wincanton and Leigh Common, there are several gentle anticlines. Beneath Wincanton, radial dips of about 4° occur in the Cornbrash. Similarly, the Cornbrash forms a half-dome structure east of Bayford.

Mere segment

The Mere Fault near Zeals (Figure 33) throws folded Corallian strata against Upper Greensand. North of the fault, undeformed Upper Greensand dips gently south-eastward at 3° [7858 3131]. South of the fault, in the hanging wall, fold axes are oriented at a large angle to the fault. Dips are up to 25° on the flanks, and, in the fold core east of Bourton, some strata are near vertical. The faulted sequence in the core of the fold east of Bourton is known from cuttings [7860 3123 to 7861 3130] along the A303 (Bristow et al., 1992, p.142). The Mere Fault, dipping 60°S, has no surface feature at this point.

Figure 31 Stratal displacement along the length of the Mere Fault plotted from depth-converted seismic data (see also Barton et al., 1998).

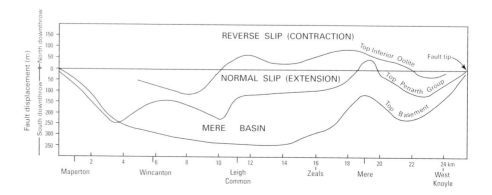

a.

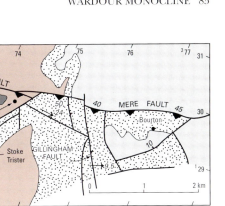
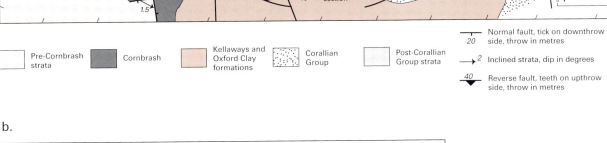

| | Pre-Cornbrash strata | | Cornbrash | | Kellaways and Oxford Clay formations | | Corallian Group | | Post-Corallian Group strata |

— 20 Normal fault, tick on downthrow side, throw in metres

→ 2 Inclined strata, dip in degrees

▼ 40 Reverse fault, teeth on upthrow side, throw in metres

b.

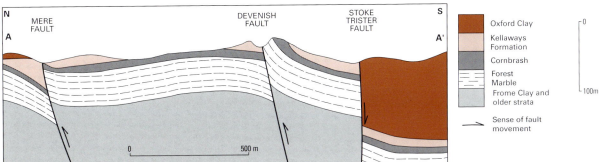

	Oxford Clay
	Kellaways Formation
	Cornbrash
	Forest Marble
	Frome Clay and older strata
	Sense of fault movement

Figure 32   Wincanton segment of the Mere Fault showing the offset west of Wincanton.

Cross-section through the Wincanton segment of the Mere Fault (line of section of A–A' shown on Figure 32a). Both the Mere and Devenish faults are interpreted as reactivated Neogene structures, whereas the Stoke Trister Fault is an early extensional structure. Vertical exaggeration ×4.

Just to the north [7856 3134], a sliver of Lower Chalk is preserved, with a faulted southern boundary against the Upper Greensand and a conformable northern boundary.

A cross-section parallel to the fault (Figure 33b) shows two broadly symmetrical anticlines with fold limbs inclined at about 10° and south-plunging fold axes about 1.5 km apart, separated by a broad syncline (Barton et al., 1998). Dips between 20° and 30° occur on the faulted outcrop [800 318] of Clavellata Beds east of Zeals House (Barton, 1994; Mottram, 1957, p.163; 1961, p.195).

The fault at Mere is a single fracture that throws Cretaceous strata down to the north against Kimmeridge Clay. The fault has a sinuous trace with an overall east-north-east trend along a 6 km-long fault portion east of Zeals House [796 319]; for part of this distance, the fault is inclined at 70° south. North of the fault, strata dip eastward or south-eastward at 1 to 2° giving rise to long dip slopes of Upper Greensand and a wide crop of Lower Chalk. South of the fault, Kimmeridge Clay forms the hanging wall and dips are largely unknown. Downthrows (Figure 30) are minimum estimates in which pre-Cretaceous and post-Creta-

ceous erosion along the segment are not separated. The northerly downthrow is 85 to 90 m between Zeals House and Mere, and increasing to more than 120 m at West Knoyle. The minimum reverse displacement downthrows Upper Chalk on the north side against the upper part of the Lower Kimmeridge Clay. Estimates of true reverse displacement along the Mere segment exceed 200 m (see Barton et al., 1998 for a full discussion).

WARDOUR MONOCLINE

At West Knoyle, the Mere Fault swings slightly northward for about 800 m along a head-filled valley; beyond this valley it cannot be recognised as a surface fracture. The reverse throw is at least 120 m at West Knoyle. East and south of the fault, strata between the Gault and the lower part of the Upper Chalk are included in a discontinuous belt of steep, northerly dips. The strata form the northern limb of an asymmetric fold, here termed the Wardour Monocline (the 'Wardour Anticline' of Edmunds, 1938),

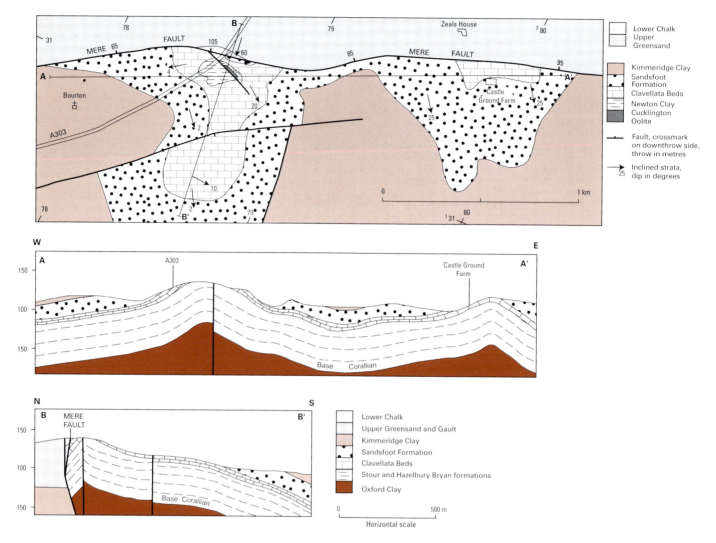

Figure 33   Zeals segment of the Mere Fault showing inliers of the Corallian Group.

Cross-section through the Zeals segment of the Mere Fault (line of sections A–A′, B–B′ shown on Figure 33a). Vertical exaggeration ×4. The apparent downthrow across the fault is at a maximum adjacent to anticlinal fold axes. (for location see Figure 30).

the anticlinal axis of which is located close to the southern edge of the steep belt (Figure 30). The fold is analogous to monoclinal folds of the type that are associated with the growth of reverse faults.

Seismic data reveal a concealed reverse fault, parallel and *en échelon* to the Mere Fault, but offset about 3 km to the south. This displaces the Top Penarth Group and Top Inferior Oolite reflectors by 30 to 75 m down to the north. This fault underlies a 6 km-long eastern segment of the Wardour Monocline, between East Knoyle and

Fonthill Bishop (Figure 30). The monocline also has a short western segment between West Knoyle and Milton which strikes approximately south-east.

Other folds in the district are much smaller and less extensive than the Wardour Monocline. Just south of Zeals, a north-trending fold exposes Newton Clay and Clavellata Beds in its core [around 785 312] (described above). Towards the eastern margin of the district, there is a shallow, south-east-trending syncline in the Upper Greensand [916 308].

NINE

Quaternary

The Quaternary deposits of the district reflect oscillations of climate, ranging from cold periglacial to warm temperate. Clay-with-flints may have formed during one of the warm periods. During the more extreme cold periods, ice sheets pushed southwards across England, but they did not reach the district. Nevertheless, it is probable that during these cold periods there were semi-permanent ice caps on the chalk hills. Marked changes in sea level were associated with these climatic oscillations. Meltwater rivers from the impermanent ice caps were graded to base levels that were at times considerably above or below that of today. These rivers carried large volumes of sand and gravel, which were deposited as a suite of fluvial deposits preserved as river terrace deposits. Periglacial conditions, which extended southwards beyond the ice sheet, were responsible for the formation of older head and head deposits. Because of the lack of fossils or radiometrically dated material, it is not possible to relate these older Quaternary deposits to the sequence of named stages recognised elsewhere in the British Isles. During the Flandrian, the most recent of the Quaternary deposits, the alluvium, was laid down.

CLAY-WITH-FLINTS

The clay-with-flints of the district is of two types. The first consists dominantly of brown clay with angular flints and corresponds to clay-with-flints as defined by Barrow (1919). The second consists of dense, clast-supported gravel with a sparse brown to reddish brown clay matrix. Because the two types are gradational, they cannot be mapped as separate units (see below). In addition, mapping the edge of the gravelly deposits is difficult, because a wash of gravel trails downslope from it.

Clay-with-flints of the first type occurs mainly in the east of the district along Great Ridge [around 930 365]. Smaller spreads occur on the crest between Parsonage Down [892 387] and just north of West Knoyle [around 865 331]. The base of the deposit is locally very gravelly [e.g. 891 388 and around 905 355] and was worked for gravel. The more gravelly deposits occur mainly on the ridge north of Charnage [850 350] and Pertwood downs [870 380 to 890 375]. East of the last point, the flinty deposits pass into more clay-rich deposits.

The clay-with-flints in the district is developed at altitudes above 200 m OD and is generally higher in the west, over 275 m OD on Brimsdown Hill [825 392], on the Seaford Chalk surface. The deposits vary in thickness from less than 1m to about 3 m.

The origin of the clay-with-flints is uncertain. Hull and Whitaker (1861) noted that it occurs only on the Chalk surface. Later (Whitaker, 1864) concluded that the deposit is of many ages, that it may be forming at the present

day, and that it was due largely to the slow decomposition of the Chalk. However, Reid (1899) pointed out that an excessive thickness of Chalk would need to be dissolved in order to produce sufficient clay residue to form clay-with-flints, and that the bulk of the material was derived from Tertiary rocks. Wilson et al. (1958) suggested that the more gravelly deposits result from removal of the matrix by water. Loveday (1962) proposed that clay from an overlying formation moved downwards by colloidal suspension to be deposited at the top of the chalk because of the filtering effect of the chalk or a change in the chemical conditions at the junction. If the overlying formation is dominantly clay, then the downward movement can be purely mechanical, as suggested Hodgson et al. (1967). These authors also suggest that dissolution of the chalk and cryoturbation took place. They proposed a cyclical development during the middle and late Pleistocene, with solution of the chalk and downward migration of the clay taking place in warm interglacial periods, and mixing by cryoturbation taking place in the glacial intervals.

OLDER HEAD

Patches of mottled orange and grey, sandy and gravelly clay, up to 2.2 m thick, and resting on up to 0.5 m of clayey gravel, occupy terrace-like features capping interfluves and small hills in the east of the district. Although the tops of the deposits are planar, they appear to be unrelated to the present-day river system and are regarded as dissected remnants of a solifluction sheet. The surface of the head between Gillingham and Motcombe falls gently west-north-westwards from 90 m to about 70 m OD. This spread is thought to be related to a solifluction sheet extending from the Upper Greensand escarpment east of Motcombe. Northwards, near Mere, the deposits appear to pass into undifferentiated, high-level, river terrace deposits. The clasts are mostly angular and subangular chert and sandstone derived from the Upper Greensand, but do include some flint. Cherty clay with clasts, presumed to be from the Portland Stone, rests on the Portland Stone near Ruddlemoor Farm [903 304], Newton [around 910 290] and near Fonthill Gifford [930 305 and 932 310]; they may be part of solifluction sheets, or possibly they are dissolution products of the underlying limestone, akin to clay-with-flints.

HEAD

Head occurs in the valley bottoms and to lesser extents on the valley slopes. The slope deposits, although locally extensive, have an irregular distribution and are difficult to map out; locally, as for example south of Fonthill Gifford [around 927 307], they are related to springlines. Deposits on the Upper Greensand have only been distinguished along the foot of the escarpment north-east of Horningsham [northwards from 821 428]. They consist mostly of chert clasts in a sand matrix and are essentially scree deposits; their maximum thickness is unknown. In Witham Park, an exposure [7718 3969] showed 2 m of clast-supported chert nodules, up to 20 cm across, in a sparse sandy matrix.

Head deposits result from the downhill movement (solifluction) and accumulation of weathered and unconsolidated rock debris under periglacial freeze-thaw conditions. Some deposits mapped as head, particularly those associated with springlines, are of colluvial origin and are probably still forming at the present day. Some of the principal spreads along the dry valleys on the Chalk probably include waterlaid material. There is commonly a step in the surface of the head deposits where those along a tributary valley join the main valley [e.g. 8872 3966 and 8872 3962]. A similar feature has been noted in the Shaftesbury area (Bristow et al., 1995).

The deposits are heterogeneous and vary in composition, but broadly reflect the upslope parent material. They commonly consist of a clayey upper unit and a gravelly base. Generally, deposits near to the Chalk outcrop consist of angular and subangular flints in a sandy clay matrix. A representative borehole [8833 3072] at East Knoyle proved, in downward succession, 0.8 m of chalk and flint gravel: 0.5 m of silt with some chalk fragments: 1.4 m of silty clay with some flint gravel: 0.2 m of silty sandy clay with some flint and chert fragments resting on Upper Greensand; total thickness is 2.9 m.

Head in the dry valleys crossing the Chalk usually has a gravelly clay top, overlying chalk gravel. In the Shaftesbury district, exposures on a valley side showed 1.5 m of stratified chalk rubble with scattered flints dipping parallel to the hillside (Bristow et al., 1995). Such deposits are probably common, but without exposure are difficult to distinguish from in situ chalk. In south-east England, deposits of chalky head are referred to as Coombe Deposits; the term Coombe Rock is restricted to cemented chalky head (Smart et al., 1966).

On the older Cretaceous deposits, the clasts are commonly of Upper Greensand chert. A representative auger hole [8968 2613] at Gutch Common encountered, in downward succession, 0.5 m of stony sand: 0.8 m of glauconitic sandy clay: 0.6 m of dark grey, glauconitic sandy clay: 0.1 m of glauconitic sand: 0.3 m of peaty sandy clay. On the Jurassic outcrops, clasts of subrounded limestone and calcareous sandstone are more common than flint. Deposits formed on the slope below the Forest Marble escarpment results, in part, from the disintegration of old landslips. Deposits in these areas commonly consist of a gravel of lensoid disks of flaser sandstone and shelly biosparite slabs. Deposits up to 3 m thick occur where these solifluction sheets were trapped behind the ridge formed by the Fuller's Earth Rock. On the gentler slopes on the clay formations, particularly the Lower Fuller's Earth, there are extensive, generally thin (up to 1.4 m), spreads of brown or orange-brown, mottled gritty, pebbly clay, commonly with a basal gravel. On many east-facing slopes on

Oxford Clay, orange-brown, clayey sand or sandy clay occur and are between 1.5 to 3 m, thick. They contain pockets of glauconite and subangular to subrounded, coarse (1 to 10 cm across) chert clasts that are particularly common near the base [e.g. around 760 405]. A representative section [7607 4090] at Walk Farm showed up to 2 m of reddish brown, very sandy clay with pods of gravel, on a lens (up to 30 cm thick) of coarse, subrounded chert gravel, with clasts 1 to 10 cm across, in irregular pockets up to 1.5 m deep, in Oxford Clay. Such deposits may represent remnants of a larger sheet of head spreading from the base of the Corallian/Cretaceous escarpment.

Head is mapped as lens-shaped bodies along the Mere Fault, and may extend as a narrow deposit, up to 2.1 m thick, along the whole length. West of Mere, the deposits are about 100 m wide [805 320]; at West Hill Farm [8435 3274], east of Mere, the head consists of flint cobbles.

A maximum thickness of 4.5 m was proved in a borehole [7505 2992] south-west of Bourton, where 1.8 m of silty sandy clay rests on 2.7 m of cobble gravel with limestone clasts. In a borehole [9150 2499] at Donhead St Andrew, 2.5 m of clay and silt passes down into 1 m of flint and chert gravel in a silty clay matrix, resting on Lower Greensand.

RIVER TERRACE DEPOSITS

River terrace deposits occur at four principal levels in the district. In addition, some high-level spreads, particularly south of Mere, remain unclassified. The numbered terraces range in height from 0.5 to 20 m above the alluvial plain. Local numbering systems have been adopted for each catchment area, as it is difficult to relate the terrace systems of each area; for example, deposits numbered as Second terrace along the River Sem lie some 5 to 6 m above the floodplain, whereas similar numbered terraces along the River Wylye are 2 to 4 m above the floodplain. In general, the height of the terrace deposits above the floodplain is about 12 to 15 m for the Fourth, 6 to 9 m for the Third, 2 to 6 m for the Second, and 0.5 to 2 m for the First. First and Second terraces occur throughout the river valleys, but higher terraces are confined to the River Wylye in the north-east. Undifferentiated terraces form flat-topped or gently south-inclined caps to interfluves or hilltops on the Kimmeridge Clay south of Mere, where they occur at two levels, about 140 m and 110 m above OD, and up to 20 to 25 m above the floodplain. Terraces in the south, between White Hill and Barrow Street, are sub-horizontal, but deposits farther north dip southwards at about 1° away from the Chalk escarpment.

The deposits consist mainly of an upper unit (up to 2.6 m thick) of sandy clay resting on gravel (0.1 to 0.5 m thick). The clay contains scattered clasts of subangular flint, chert and, less commonly, sandstone. The gravel is clayey and poorly sorted. Clasts consist of flint, chert and sandstone with some limestone and chalk; they are generally less than 6 cm across, but some are up to 12 cm. Deposits in the Frome catchment area are more sandy and are either sandy gravel or clayey sandy gravel, with clasts consisting of coarse angular chert. The deposits along the

River Cale south of Wincanton consist mainly of up to 2.5 m of very sandy, orange-brown clay or clayey sand, resting on 0.1 to 0.5 m of clayey sandy gravel with clasts, 0.5 to 2 cm across, mainly of Forest Marble or Cornbrash limestone. The gravelly base of the undifferentiated terraces adjacent to and up to 1 km south of the Mere Fault consists of a mixture of flint and small fragments of hard or nodular chalk. Farther south, it consists of subrounded flints in a silty clay matrix. Terraces Three and Four appear to be more gravelly than lower ones. Because of their high clay content, the terraces have rarely been worked for gravel. An exception is at Knook, where there are small workings [940 415] in Second terrace deposits.

Thicknesses are unknown for most spreads. Second terrace deposits are at least 2.8 m thick west of Heytesbury [9130 4287]. Second terrace deposits bordering the Shreen Water and River Lodden vary in thickness from 0.5 to 1.4 m of sandy clay, on 0.1 to 0.4 m of gravel. Deposits along the River Cale comprise an upper unit, up to 2.5 m thick, on 0.3 to 0.5 m of gravel. Undifferentiated terrace deposits from Mere southwards, consist of 0.5 to 2.1 m of sandy clay on 0.1 to 0.5 m of gravel.

ALLUVIUM

Alluvium is the floodplain deposit of the modern river system; it is most extensive along the River Wylye, and also occurs along its major tributaries and other rivers in the district. It generally consists of an upper unit of mottled grey and brown, or orange-brown, commonly organic or peaty, silt, silty clay and clayey sand with only scattered clasts, and a lower unit of sand and gravel, with a clay matrix in places. The thickness of the two units varies from 0.7 to 1.9 m and up to 1.4 m respectively. Locally, the two units are not readily separable; in places, the alluvium consists of up to 1.6 m of pebbly sandy clay [e.g. 8568 2505]; elsewhere [for example 870 406] up to 3 m of flint gravel were proved.

The drainage of the Bow Brook in the south of the district is restricted to the west side of the well-featured outcrop of the Peterborough Member, with the drainage from the Forest Marble–Cornbrash dip slope forming wide alluvial flats extending south-eastwards across the Kellaways Formation from Horsington Marsh [7120 2470] to beyond the district margin [7250 2380]. The floodplains of the Bow Brook and River Cale merge where the Bow Brook is able to pass eastwards across the Peterborough Member where it is breached by faulting.

Tufa

Along the valley sides south of Higher Alham [around 6793 4089 and 6790 4059], deposits of white-weathering, crumbly, porous tufa is associated with springs emerging from the bases of the Junction Bed and Pennard Sands.

INFORMATION SOURCES

Further geological information held by the British Geological Survey relevant to the Wincanton district is listed below. It includes published material in the form of maps, memoirs and reports, and unpublished maps and reports. Also included are other sources of data held by BGS in a number of collections, including borehole records, fossils, rock samples, thin sections, clay mineralogy, grain size, hydrogeological and bibliographical data, and photographs.

Searches of indices to some of the collections can be made on the Geoscience Index System in BGS libraries. This is a developing computer-based system which carries out searches of indices to collections and digital databases for specified geographical areas. It is based on a geographical information system linked to a relational database management system. Results of the searches are displayed on maps on the screen. At the present time (1998), the datasets are limited and not all are complete. The indices which are available are listed below:

- Index of boreholes
- Outlines of BGS maps at 1:50 000 and 1:10 000 scale and 1:10 560 scale County Series
- Geochemical sample locations on land
- Aeromagnetic and gravity data recording stations

Maps

GEOLOGY MAPS:

1:1 000 000
Pre-Permian geology of the United Kingdom (south), 1985

1:625 000
Solid geology (south sheet)
Quaternary geology (south sheet)

1:584 000
Tectonic map of Great Britain and Northern Ireland

1:250 000
Bristol Channel (solid geology), 1988, (solid and drift), 1986
Portland (solid geology), 1983

1:63 360 (Old Series)
The district was first geologically surveyed on the one-inch (1:63 360) scale by H W Bristow and published on Old series Sheets 15, 18 and 19 respectively in 1856, 1850 and 1845, and by W T Aveline, published on Old series Sheet 14 in 1857. A revised edition of Sheet 14 was issued in 1859. J H Blake, H B Woodward and W A Ussher carried out minor revisions of the Jurassic strata on sheets 18 and 19 between 1867 and 1871, and the maps were reissued respectively in 1875 and 1873. A third edition of Sheet 19, incorporating work by A J Jukes-Browne, was published in 1899.

1:50 000 and 1:63 360 (New Series)
Sheet 280 (Wells) Solid and Drift, 1963, reprinted (at 1:50 000 scale), 1984
Sheet 281 (Frome) Solid and Drift, 1965, revised edition (at 1:50 000 scale) reprinted 1998
Sheet 282 (Devizes) Solid and Drift, 1899, reprinted 1985
Sheet 296 (Glastonbury) Solid and Drift, 1969, reprinted (at 1:50 000 scale), 1973

Sheet 297 (Wincanton) Solid and Drift, 1996
Sheet 298 (Salisbury) Solid and Drift, 1903, reprinted (at 1:50 000 scale), 1976
Sheet 312 (Yeovil) Solid and Drift, 1958, reprinted (at 1:50 000 scale), 1973
Sheet 313 (Shaftesbury), Solid and Drift, 1994
Sheet 314 (Ringwood), Solid and Drift, 1902, reprinted (at 1:50 000 scale), 1976

1:10 000 and 1:10 560
Revision of the Cretaceous mapping at the six-inch (1:10 560) scale on County sheets was carried out by A J Jukes-Browne between 1889 and 1900, by F J Bennett between 1894 and 1896, and by C Reid between 1895 and 1900. The maps are not published, but are available for consultation in the BGS Libraries at Keyworth, Exeter and London.

1:10 560 scale Wiltshire County Geological Sheets
51 (SW and SE) A J Jukes-Browne, 1889; partial resurvey by D R A Ponsford in 1956.
52 (SW and SE) A J Jukes-Browne, 1894; partial resurvey by D R A Ponsford in 1956.
57 (NW, part only) A J Jukes-Browne, 1894; (NE, part only) A J Jukes-Browne, 1894 and F J Bennett, 1896; (SW, part only) A J Jukes-Browne, 1894.
58 (NW, NE, SW and SE) F J Bennett between 1894 and 1896.
63 (NE and SE, part only) A J Jukes-Browne, 1890.
64 (NW, part only) A J Jukes-Browne, 1890, (NE) A J Jukes-Browne, 1890, (SW) A J Jukes-Browne, 1890 and C Reid, 1895, (SE) C Reid, 1900.
69 (NW, part only) C Reid, 1900, (NE) C Reid, 1898).

1:10 000 scale National Grid Geological Sheets
Maps at the 1:10 000 scale covering the whole of 1:50 000 Sheet 297 are listed below, together with the surveyors' initials and dates of the survey. Data surveyed on 1:10 560 scale County sheets (Somerset and Wiltshire) for the areas approximately north of grid line 43 (falling on the Frome (281) Sheet) have been reconstituted at the 1:10 000 scale on to the National Grid sheets. The surveyors were: C M Barton, C R Bristow, E C Freshney, D R A Ponsford, R T Taylor, F B A Welch and R K Westhead. Copies are available for consultation in the BGS Libraries at Keyworth, Exeter and the London Information Office. Dyeline copies can be purchased from the Sales Desk.

ST 62 NE	(Holton)	ECF	1992
ST 62 SE	(Charlton Horethorne)	RTT	1986–1988
ST 63 NE	(Batcombe)	RKW	1992
ST 63 SE	(Bruton)	ECF	1992
ST 64 SE	(Chesterblade)	RKW, DRAP, FBAW	1993, 1954–1955
ST 72 NW	(Wincanton)	ECF	1994
ST 72 NE	(Cucklington)	CRB	1989
ST 72 SW	(Templecombe)	RTT	1989–1990
ST 72 SE	(West Stour)	CRB	1988
ST 73 NW	(Brewham)	RKW	1994
ST 73 NE	(Kilmington)	ECF, CMB, CRB, RKW	1994
ST 73 SW	(Charlton Musgrove)	ECF	1993
ST 73 SE	(Zeals)	ECF	1994
ST 74 SW	(Wanstrow)	RKW, FBAW	1993, 1954–1955

ST 74 SE	(Trudoxhill)	RKW, FBAW	1994, 1956
ST 82 NW	(Gillingham)	CRB	1992
ST 82 NE	(Semley)	CRB	1989 and 1992
ST 82 SW	(East Stour)	CRB	1987
ST 82 SE	(Shaftesbury)	CRB	1988
ST 83 NW	(Maiden Bradley)	CMB, CRB	1994
ST 83 NE	(Brixton Deverill)	CRB	1994
ST 83 SW	(Mere)	CMB	1994
ST 83 SE	(East Knoyle)	CRB	1994
ST 84 SW	(Horningsham)	CRB, DRAP	1994, 1956
ST 84 SE	(Longbridge Deverill)	CRB, DRAP	1994, 1956
ST 92 NW	(Tisbury)	CRB	1989 and 1992
ST 92 SW	(Berwick St John)	CRB	1988
ST 93 NW	(Great Ridge)	CRB	1993
ST 93 SW	(Hindon)	CRB	1993
ST 94 SW	(Heytesbury)	CRB	1993, 1956

GEOPHYSICAL MAPS

1:1 500 000
Colour shaded relief gravity map of Britain, Ireland and adjacent areas, 1997
Colour shaded relief magnetic map of Britain, Ireland and adjacent areas, 1998

1:250 000
Bristol Channel (Sheet 51° N–04°W)
  Aeromagnetic, 1980
  Gravity, 1986
Portland (Sheet 50° N–04°W)
  Aeromagnetic, 1978
  Gravity, 1978

1:50 000
Geophysical information maps; these are plot-on-demand maps which summarise graphically the publicly available geophysical information held for the sheet in the BGS databases. Features include:

• Regional gravity data: Bouguer anomaly contours and location of observations.
• Regional aeromagnetic data: total field anomaly contours and location of digitised data points along the flight lines.
• Gravity and magnetic fields plotted on the same base map at 1:50 000 scale to show correlation between anomalies.
• Separate colour contour plots of gravity and magnetic fields at 1:125 000 scale for easy visualisation of important anomalies.
• Location of local geophysical surveys.
• Location of public domain seismic reflection and refraction surveys.
• Location of deep boreholes and those with geophysical logs.

HYDROGEOLOGICAL MAP

1:100 000
Hydrogeological map of the Chalk and associated minor aquifers of Wessex, 1979.

Books

Memoirs, books, reports and papers relevant to the Wincanton district arranged by topic. Some are out of print or are not widely available, but may be consulted at BGS and other libraries.

GENERAL GEOLOGY

British regional geology

GREEN, G W. 1992. *British regional geology, Bristol and Gloucester region (third edition).* (London: HMSO for British Geological Survey.)

KELLAWAY, G A, and WELCH, F B A. 1948. *British regional geology: Bristol and Gloucester district (2nd edition).* (London: HMSO for Institute of Geological Sciences.)

MELVILLE, R V, and FRESHNEY, E C. 1982. *British regional geology: Hampshire Basin and adjoining areas.* (London: HMSO for Institute of Geological Sciences.)

WELCH, F B A, and CROOKALL, R. 1935. *British regional geology. Bristol and Gloucester District.* (London: HMSO for British Geological Survey.)

Sheet memoirs

BRISTOW, C R, BARTON, C M, FRESHNEY, E C, WOOD, C J, EVANS, D J, COX, B M, IVIMEY–COOK, H C, and TAYLOR, R T. 1995. Geology of the country around Shaftesbury. *Memoir of the Geological Survey of Great Britain*, Sheet 313 (England and Wales).

GREEN, G W, and WELCH, F B A. 1965. Geology of the country around Wells and Cheddar. *Memoir of the Geological Survey of Great Britain*, Sheet 280 (England and Wales).

JUKES–BROWNE, A J. 1905. The geology of the country south and east of Devizes. *Memoir of the Geological Survey of Great Britain*, Sheet 282 (England and Wales).

REID, C. 1902. The geology of the country around Ringwood. *Memoir of the Geological Survey of Great Britain*, Sheet 314 (England and Wales).

REID, C. 1903. Geology of the country around Salisbury. *Memoir of the Geological Survey of Great Britain*, Sheet 298 (England and Wales).

WILSON, V, WELCH, F B A, ROBBIE, J A, and GREEN, G W. 1958. Geology of the country around Bridport and Yeovil. *Memoir of the Geological Survey of Great Britain*, Sheet 312 and 327 (England and Wales).

Memoirs and other books

DE LA BECHE, H T. 1846. On the formation of the rocks of south Wales and south western England. *Memoir of the Geological Survey of Great Britain*, Vol. 1, 1–296.

EDWARDS, W, and PRINGLE, J. 1926. On a borehole in the Lower Oolitic Rocks at Wincanton, Somerset. *Summary of Progress of the Geological Survey for 1925*, 183–188.

JUKES-BROWNE, A J, and HILL, W. 1900. The Cretaceous rocks of Britain. 1. The Gault and Upper Greensand. *Memoir of the Geological Survey of Great Britain.*

JUKES-BROWNE, A J, and HILL, W. 1903. The Cretaceous rocks of Britain. 2. The Lower and Middle Chalk of England. *Memoir of the Geological Survey of Great Britain.*

JUKES-BROWNE, A J, and HILL, W. 1904. The Cretaceous rocks of Britain. 4. Upper Chalk of England. *Memoir of the Geological Survey of Great Britain.*

PRINGLE, J. 1909. On a boring in the Fullonian and Inferior Oolite at Stowell, Somerset. *Summary of Progress of the Geological Survey of Great Britain for 1908*, 83–86.

PRINGLE, J. 1910. On a boring at Stowell, Somerset. *Geological Survey of Great Britain, Summary of Progress for 1909*, 68–70.

WHITTAKER, A (editor). 1985. *Atlas of onshore sedimentary basins in England and Wales.* (Glasgow: Blackie.)

WOODWARD, H B. 1893. The Jurassic rocks of Britain, Vol. 3. The Lias of England and Wales (Yorkshire excepted). *Memoir of the Geological Survey of Great Britain.*

WOODWARD, H B. 1894. The Jurassic rocks of Britain. IV. The Lower Oolitic rocks of England. *Memoir of the Geological Survey of Great Britain.*

WOODWARD, H B. 1895. The Jurassic rocks of Britain. Vol. 5. The Middle and Upper Oolitic rocks of England (Yorkshire excepted). *Memoir of the Geological Survey of Great Britain.*

Reports

RESEARCH REPORTS

BARTON, C M, IVIMEY-COOK, H C, LOTT, G K, and TAYLOR, R T. 1993. The Purse Caundle Borehole, Dorset: Stratigraphy and sedimentology of the Inferior Oolite and Fuller's Earth in the Sherborne area of the Wessex Basin. *Research Report of the British Geological Survey, Onshore Geology Series,* SA/93/01.

CHADWICK, R A, and KIRBY, G A. 1982. The geology beneath the Lower Greensand/Gault surface in the Vale of Wardour area. *Report of the Institute of Geological Sciences,* 81/1, pp.15–18.

DOWNING, R A, and PENN, I E. 1992. Groundwater flow during the development of the Wessex Basin and its bearing on hydrocarbon and mineral resources. *British Geological Survey Research Report,* SD/91/1.

HOLLOWAY, S. 1982. Bruton No.1. Geological Well Completion Report. *Institute of Geological Sciences, Deep Geology Unit,* Report No. 82/8.

TECHNICAL REPORTS

The Technical reports listed below are detailed accounts of the geology of the constituent 1:10 000 scale maps of the Wincanton 1:50 000 Series Sheet 297. Copies of the reports may be ordered from the British Geological Survey, Keyworth and Exeter.

ST 62 SE   TAYLOR, R T. 1990. Geology of the Charlton Horethorne district (Somerset and Dorset). *British Geological Survey Technical Report,* WA/89/69.

ST63SE and ST62NE   FRESHNEY, E C. 1994. Geology of the Bruton–Holton district, Somerset. *British Geological Survey Technical Report,* WA/94/02.

ST63NW and ST63NE   BRISTOW, C R, and WESTHEAD, R K. 1993. Geology of the Evercreech–Batcombe district (Somerset). *British Geological Survey Technical Report,* WA/93/89.

ST64SE and ST74SW   WESTHEAD, R K. 1994. Geology of the Chesterblade–Wanstrow district (Somerset). *British Geological Survey Technical Report,* WA/93/90.

ST72NE and ST72SE   BRISTOW, C R. 1990a. Geology of the West Stour–Cucklington district. *British Geological Survey Technical Report,* WA/90/48.

ST72SW   TAYLOR, R T. 1991. Geology of the Templecombe district (Somerset). *British Geological Survey Technical Report,* WA/91/31.

ST72NW and ST73SW   FRESHNEY, E C. 1995a. Geology of the Charlton Musgrove–Wincanton district, Somerset. *British Geological Survey Technical Report,* WA/95/23.

ST73NE and ST73SE   FRESHNEY, E C. 1995b. Geology of the Kilmington–Zeals district, Wiltshire. *British Geological Survey Technical Report,* WA/95/24.

ST73NW and ST74SE   WESTHEAD, R K. 1995. Geology of the Brewham–Trudoxhill district (Somerset and Wiltshire). *British Geological Survey Technical Report,* WA/95/10.

ST82NW and ST82NE   BRISTOW, C R. 1993. Geology of the Gillingham–Semley district, Dorset. *British Geological Survey Technical Report,* WA/93/87.

ST82SW and ST82SE   BRISTOW, C R. 1989. Geology of the East Stour–Shaftesbury district (Dorset). *British Geological Survey Technical Report,* WA/89/58.

ST83SW and ST83NW   BARTON, C M. 1994. Geology of the Mere–Maiden Bradley district (Wiltshire and Dorset). *British Geological Survey Technical Report,* WA/94/87.

ST83NE and ST83SE   BRISTOW, C R. 1995a. Geology of the Brixton Deverill–East Knoyle district (Wiltshire). *British Geological Survey Technical Report,* WA/95/13.

ST84SW and ST84SE   BRISTOW, C R. 1994. Geology of the Horningsham–Warminster south district (Wiltshire). *British Geological Survey Technical Report,* WA/94/88.

ST92SW   BRISTOW, C R. 1990b. Geology of the Berwick St John district (Wiltshire). *British Geological Survey Technical Report,* WA/90/49.

ST92NW and part of ST93SW   BRISTOW, C R. 1995b. Geology of the Tisbury district (Wiltshire). *British Geological Survey Technical Report,* WA/95/82.

ST93NW and ST93SW   BRISTOW, C R. 1995c. Geology of the Great Ridge–Hindon district (Wiltshire). *British Geological Survey Technical Report,* WA/95/85.

ST94SW   BRISTOW, C R. 1996. Geology of the Heytesbury district, Wiltshire. *British Geological Survey Technical Report,* WH/95/86.

BIOSTRATIGRAPHY

WOODS, M A, and BRISTOW, C R. 1995. A biostratigraphical review of the Gault, Upper Greensand and Chalk of the Wincanton (297) district, Wiltshire. *British Geological Survey Technical Report,* WA/95/60.

BULK MINERALS

GALLOIS, R W. 1979. *Oil shale resources in Great Britain.* (London: Institute of Geological Sciences.)

ENGINEERING GEOLOGY

GOSTELOW, T P. 1991. Geological processes and their effect on the engineering behaviour of the Gault–UGS: An example from Shaftesbury, Dorset. *British Geological Survey Technical Report,* WN/91/9.

GEOPHYSICS

CHACKSFIELD, B C. 1997. Interpretation of magnetic anomalies in the Wincanton District. *British Geological Survey, Regional Geophysics,* PN/95/2.

CORNWELL, J D, and CHACKSFIELD, B C. 1997. Geophysical investigations in the Wincanton district. *British Geological Survey Technical Report,* WK/96/10.

PETROGRAPHY AND CLAY MINERALOGY

BRISTOW, C R, and LOTT, G K. 1994. The stratigraphy and building stone potential of the Portland Beds in the western part of the Vale of Wardour. *British Geological Survey Technical Report,* WA/94/97C.

LOTT, G K. 1993. Thin section petrography of Upper Jurassic (Corallian Oolite Fm/Kimmeridge Clay Fm/Portland 'Stone') sediments in the Salisbury Sheet (298) area, Wiltshire. *British Geological Survey Technical Report,* WH/93/81.

LOTT, G K. 1994. The petrology of sandstones and siltstones from the Portland Sand Formation, Tisbury Borehole, Wiltshire. *British Geological Survey Technical Report*, WH/94/178R.

MITCHELL, C J and MURPHY, H A. 1995. Clay mineralogy of Frome Clay from Lodge Farm and Bratton Seymour, Somerset. *British Geological Survey Technical Report*, MPSR/95/21.

RADON

APPLETON, J D, and BALL, T K. 1995. Radon and background radioactivity from natural sources: characteristics, extent and relevance to planning and development in Great Britain. *British Geological Survey Technical Report*, WP/95/2.

WATER SUPPLY

RICHARDSON, L. 1928. Wells and springs of Somerset. *Memoir of the Geological Survey of Great Britain*.

WHITAKER, W, and EDMUNDS, F H. 1925. The water supply of Wiltshire from underground sources. *Memoir of the Geological Survey of Great Britain*.

WHITAKER, W, and EDWARDS, W. 1926. Wells and springs of Dorset. *Memoir of the Geological Survey of Great Britain*.

Documentary collections

BOREHOLE RECORD COLLECTION

BGS holds collections of records of boreholes which can be consulted at BGS, Exeter, where copies of most records can be purchased. For the Wincanton district, the collection consists of the sites and logs of about 770 boreholes. Index information, which includes site references, for these bores has been digitised. The logs are either handwritten or typed, and many of the older records are drillers' logs. The more important boreholes, for which there are either cores or geophysical logs (in some cases both), are listed below.

Important boreholes in the Wincanton and immediately adjacent district

Name	Grid reference	BGS Regist. No.	Data held	Purpose
Alham No. 7	ST 6793 4118	ST64SE/6	1	Water supply
Bayford	ST 7287 2884	ST62NE/39	2	Water supply
Brixton Deverill	ST 8590 3834	ST83NE/4	2	Water supply
Bratton Seymour	ST 6763 2860	ST62NE/40	1, 2	Stratigraphical test
Bruton	ST 6896 3284	ST63SE/19	2	Stratigraphical test
Combe Throop	ST 7260 2350	ST72SW/1	1, 2	Stratigraphical test
Cranborne	SU 0341 0907	SU00NW/1	2	Hydrocarbon exploration
Devizes	ST 9603 5699	ST95NE/1	2	Hydrocarbon exploration
East Stour	ST 8013 2297	ST82SW/9	1, 2	Stratigraphical test
Fifehead Magdalen	ST 7985 2100	ST72SE/1	2	Hydrocarbon exploration
Fonthill Bishop	ST 9151 3517	ST93NW/2	1, 2	Water supply
Fonthill Bishop	ST 9353 3448	ST93SW/7	1, 2	Water supply
Fonthill Bishop	ST 9299 3424	ST93SW/8	1, 2	Water supply
Fonthill Bishop	ST 941 342	ST93SW/15	2	Water supply
Fonthill Bishop	ST 9471 3417	ST93SW/16	2	Water supply
Fonthill Bishop	ST 9389 3356	ST93SW/17	2	Water supply
Frome	ST 7632 4769	ST74NE/3	1	Site investigation
Gillingham	ST 7959 2661	ST72NE/4		Water supply
Holton	ST 6986 2772	ST62NE/15		Site investigation
Henstridge	ST 7243 1989	ST71NW/6		Water
Heytesbury	ST 9301 4195	ST94SW/44		Water supply
Lodge Farm	ST 6893 4133	ST64SE/18	1, 2	Stratigraphical test
Lower Pertwood	ST 8829 3652	ST83NE/7		Water supply
Monkton Deverill	ST 8566 3802	ST83NE/5	2	Water supply
Norton Ferris	ST 7820 3700	ST73NE/1	2	Hydrocarbon exploration
Oborne	ST 6520 1830	ST61NE/1	1	Water supply
Rudge	ST 8229 5232	ST85SW/1	1	Water supply
Purse Caundle	ST 7012 1826	ST71NW/7	1, 2, 3	Stratigraphical test
Shrewton	SU 0314 4199	SU04SW/1	2	Hydrocarbon exploration
Stowell	ST 6849 2180	ST62SE/1	1	Water supply
Templecombe	ST 7100 2279	ST72SW/1	1, 2	Water supply
Tisbury	ST 9359 2907	ST93NW/2	1, 2, 3	Stratigraphical test
Tytherington	ST 9191 4061	ST94SW/26		Water supply
Urchfont	SU 0444 5816	SU05NW/18	2	Hydrocarbon exploration
Westbury	ST 8615 5195	ST85SE/1		Water supply
Wincanton	ST 7116 2816	ST72NW/55	1	Water supply
Woolston	ST 6595 2820	ST62NE/39	2	Water supply
Yarnbury	SU 0336 4105	SU04SW/5	2	Hydrocarbon exploration

1 Palaeontological and/or core material in BGS collections
2 Geophysical logs in BGS archives
3 Thin sections in BGS archives

SITE-INVESTIGATION REPORTS

This collection consists of site-investigation reports carried out to investigate foundation conditions prior to construction.

Material collections and other source data

GEOLOGICAL SURVEY PHOTOGRAPHS

About 65 photographs illustrating aspects of the geology of the Wincanton district are deposited for reference in the libraries at BGS, Exeter, Edinburgh, Keyworth and the Information Office, London.

The photographs, taken during 1922, 1932, 1990 and 1996, depict details of the various rocks exposed either naturally or in excavations and also some general views. A list of titles can be supplied on request. The photographs can be supplied as black and white or colour prints (black and white only for the pre-war photographs) and 2 × 2 colour transparencies, at a fixed tariff.

PETROGRAPHICAL AND BUILDING-STONE COLLECTIONS

The petrographical and building-stone collections for the district consist of 44 hand specimens and thin sections. The collections are indexed on the basis of the 1:50 000-scale geological map, but much of the earlier part of the collection cannot at the moment be searched by National Grid Reference.

GRAIN-SIZE DATA

Grain-size data for 29 samples from the arenaceous formations ranging from the Bridport Sands to the Upper Greensand are held in the BGS Exeter Office.

CLAY-MINERAL DATA

Data on the clay mineralogy of 77 samples from the Fuller's Earth, Frome Clay and Forest Marble is held by BGS, Keyworth.

PALAEONTOLOGICAL COLLECTIONS

The collections of biostratigraphical specimens are taken from surface and temporary exposures, and from boreholes throughout the district. They comprise 152 calcareous microfaunal and 34 palynological samples, and 1000s of macrofossils. The collections are working collections and are used for reference. They are not at present on a computer database. Much detail is given in the unpublished reports listed below (marked * in the text). In most cases, it should be possible to purchase copies on application to the Group Manager of the Biostratigraphy and Sedimentology Unit, Keyworth.

COX, B M. 1990a. Wincanton Sheet (297): Kimmeridge Clay of the Gillingham Brickpit and nearby localities, Dorset. *British Geological Survey Technical Report*, WH/90/100R.

COX, B M. 1990b. Wincanton Sheet (297): Corallian Group localities on 1:10 000 sheets ST72NE. *British Geological Survey Technical Report*, WH/90/214R.

COX, B M. 1990c. Corallian sections at Kingsmead Tunnel, west of Gillingham, Dorset. *British Geological Survey Technical Report*, WH/90/302R.

COX, B M. 1991a. Macrofauna and age determination of a section in the Hazelbury Bryan Formation on 1:10 000 Sheet ST72NW. *British Geological Survey Technical Report*, WH/91/253R.

COX, B M. 1991b. Combe Throop Borehole, Combe Throop, Somerset: Oxford Clay. *British Geological Survey Technical Report*, WH/91/310R.

COX, B M. 1991c. Combe Throop Borehole, Combe Throop, Somerset: Kellaways Beds. *British Geological Survey Technical Report*, WH/91/336R.

COX, B M. 1993a. Jurassic material collected from the A303 trunk road dualling (1990–1991) between Wincanton, Somerset and Mere, Wiltshire. *British Geological Survey Technical Report*, WH/93/114R.

COX, B M. 1993b. The Kimmeridge Clay at Pitts Farm, Sedgehill. *British Geological Survey Technical Report*, WH/93/064R.

COX, B M. 1993c. Portland 'Beds' localities on 1:10 000 sheets ST92NW and ST92SW. *British Geological Survey Technical Report*, WH/93/301R.

COX, B M. 1993d. Upper Jurassic localities of the Gillingham, Dorset, area. *British Geological Survey Technical Report*, WH/93/042R.

COX, B M. 1993e. Recent collecting by E C Freshney from 1:10 000 Sheet ST73SW. *British Geological Survey Technical Report*, WH/93/281R.

COX, B M. 1994a. Notes on specimens from the Alham No.7 (1946) Borehole, near Shepton Mallet, Somerset. *British Geological Survey Technical Report*, WH/94/196R.

COX, B M. 1994b. Recent collecting from the Kimmeridge Clay of 1:10 000 sheets ST83SE and SW, and ST82NE. *British Geological Survey Technical Report*, WH/94/230R.

COX, B M. 1994c. Recent collecting from 1:10 000 Sheet ST73NW. *British Geological Survey Technical Report*, WH/94/260R.

COX, B M. 1994d. Recent collecting from the Oxford Clay and Kellaways Formations of the Wincanton district. *British Geological Survey Technical Report*, WH/94/274R.

COX, B M. 1994e. Further Kimmeridge Clay material from 1:10 000 sheets ST83SE and SW. *British Geological Survey Technical Report*, WH/94/299R.

COX, B M. 1995a. Fossils from the Cornbrash and Forest Marble formations of 1:10 000 Sheet ST73SW. *British Geological Survey Technical Report*, WH/95/080R.

COX, B M. 1995b. Fossils from the Cornbrash of a pit at South Brewham, Somerset. *British Geological Survey Technical Report*, WH/95/098R.

COX, B M. 1995c. Specimens from the Oxford Clay of two localities on 1:10 000 Sheet SY67NW. *British Geological Survey Technical Report*, WH/95/186R.

COX, B M, and IVIMEY-COOK, H C. 1995. The Lodge Farm (Batcombe) and Bratton Seymour boreholes — a macro-palaeontological assessment. *British Geological Survey Technical Report*, WH/95/047R.

IVIMEY-COOK, H C. 1991. The Middle Jurassic (Cornbrash and Forest Marble) of the Combe Throop Borehole; 1" 313, Shaftesbury. *British Geological Survey Technical Report*, WH/91/301.

IVIMEY-COOK, H C. 1992a. The stratigraphy of the Wincanton Borehole (ST72NW/55); 1" Sheet 297. *British Geological Survey Technical Report*, WH/92/314R.

IVIMEY-COOK, H C. 1992b. Lower and Middle Jurassic fossils from Sheet 297 and immediately adjacent areas. Part 1. Collections made before 1992. *British Geological Survey Technical Report*, WH/92/316R.

IVIMEY-COOK, H C. 1993a. Lower Jurassic fossils from ST 63 NW; ST 64 SW; ST 52 NW and ST 54 SE, Glastonbury. *British Geological Survey Technical Report*, WA/93/104R.

IVIMEY-COOK, H C. 1993b. Middle Jurassic fossil from ST 62 and ST 63, Wincanton; presented by H C Prudden. *British Geological Survey Technical Report*, WH/93/128R.

IVIMEY-COOK, H C. 1993c. Lower and Middle Jurassic fossils from ST62NE and ST63SE. *British Geological Survey Technical Report*, WH/93/130R.

IVIMEY-COOK, H C. 1993d. Further Middle Jurassic fossils from the Wincanton Sheet (297); from ST63; ST72 and ST73. *British Geological Survey Technical Report*, WH/93/260R.

OWENS, B. 1982. Palynological report on sample from IGS Bruton Borehole. *British Geological Survey Technical Report*, PDL/82/248[R].

RIDING, J B. 1993a. A palynological investigation of the BGS Combe Throop Borehole, Somerset. *British Geological Survey Technical Report*, WH/93/19R.

RIDING, J B. 1993b. A palynological investigation of the BGS Tisbury Borehole, Wiltshire (15.90 to 125.40 m). *British Geological Survey Technical Report*, WH/93/67R.

RIDING, J B. 1993c. A palynological investigation of the Kimmeridge Clay from North Hayes Farm, Motcombe, Dorset. *British Geological Survey Technical Report*, WH/93/322R.

RIDING, J B. 1994a. A palynological investigation of a suite of Jurassic samples from the Wincanton Sheet. *British Geological Survey Technical Report*, WH/94/49R.

RIDING, J B. 1994b. Palynological investigation of a Lower Cretaceous sample from the Wincanton Sheet. *British Geological Survey Technical Report*, WH/94/63R.

RIDING, J B. 1994c. A palynological investigation of the Upper Chalk from two localities on the Wincanton Sheet. *British Geological Survey Technical Report*, WH/94/136R.

STRANK, A R E. 1982. Foraminifera from the Bruton Borehole, Somerset. 1" 297 (BLK 4250–4257). *British Geological Survey Technical Report*, PDL/82/194[R].

THOMAS, J E. 1991. Palynology of samples from the 'Aspidoceras Cementstone' from the Wincanton Sheet. *British Geological Survey Technical Report*, WH/91/356R.

WILKINSON, I P. 1978. Kimmeridgian to ?Portlandian ostracoda from the Tisbury Borehole (1" Sheet 297). *British Geological Survey Technical Report*, PDL 78/4R.

WILKINSON, I P. 1993a. Foraminifera from the 2.6–16.8 m interval of Fonthill Bishop Borehole No. WW1R. *British Geological Survey Technical Report*, WH/93/124R.

WILKINSON, I P. 1993b. Calcareous microfaunas from a suite of samples from the Wincanton Sheet. *British Geological Survey Technical Report*, WH/93/143.

WILKINSON, I P. 1993c. Foraminifera from the Upper Chalk of the Fonthill Bishop [WW5R] Borehole. *British Geological Survey Technical Report*, WH/93/124R.

WILKINSON, I P. 1993d. Callovian–Oxfordian microfossils from a suite of auger holes from the Gillingham Sheet (297). *British Geological Survey Technical Report*, WH/93/293R.

WILKINSON, I P. 1993e. Foraminifera from the Blandford Chalk near Hindon. *British Geological Survey Technical Report*, WH/93/309R.

WILKINSON, I P. 1994a. Calcareous microfaunas from the Lodge Farm Borehole. *British Geological Survey Technical Report*, WH/94/160R.

WILKINSON, I P. 1995a. A preliminary examination of the calcareous microfaunas from the Bratton Seymour Borehole. *British Geological Survey Technical Report*, WH/95/64R.

WILKINSON, I P. 1995b. A re-appraisal of problematical taxa in the Upper Chalk of 1:50K sheets [297] 314 and 328 together with their biostratigraphical implications. *British Geological Survey Technical Report*, WH/95/118R.

WILKINSON, I P. 1995c. Foraminifera from the Upper Chalk of Fonthill Bishop Borehole. *British Geological Survey Technical Report*, WH/95/165R.

WILKINSON, I P. 1995d. Ostracoda from the Kimmeridge Clay of the Tisbury Borehole. *British Geological Survey Technical Report*, WH/95/207R.

WOODS, M A. 1991. Outline palaeontology and biostratigraphy of basal Cenomanian sediments near Zeals, Wilts. *British Geological Survey Technical Report*, WH/91/291R.

WOODS, M A. 1992. The Chalk succession at Charnage Down Chalk Quarry, Wiltshire. *British Geological Survey Technical Report*, WH/92/319R.

WOODS, M A. 1993a. Cretaceous macrofaunas from boreholes and field exposures on Sheets 297 (Wincanton), 328 (Dorchester) & 329 (Bournemouth). *British Geological Survey Technical Report*, WH/93/63R.

WOODS, M A. 1993b. Upper Chalk macrofaunas from Fonthill Bishop Borehole WW5R, Wiltshire. *British Geological Survey Technical Report*, WH/93/158R.

WOODS, M A. 1993c. Chalk macrofaunas from the country around Chicklade, Wiltshire. *British Geological Survey Technical Report*, WH/93/245R.

WOODS, M A. 1993d. Identification and interpretation of Chalk macrofaunas from sheets 281 (Frome), 297 (Wincanton) & 298 (Salisbury). *British Geological Survey Technical Report*, WH/93/291R.

WOODS, M A. 1994a. A biostratigraphical review of the Gault, Upper Greensand and Chalk of the Wincanton (Sheet 297) district and adjoining areas. *British Geological Survey Technical Report*, WH/94/168R.

WOODS, M A. 1994b. Macrofaunas from the Upper Greensand and Chalk of the Wincanton (297) and Salisbury (298) sheets. *British Geological Survey Technical Report*, WH/94/222R.

WOODS, M A. 1994c. Macrofaunas from the Upper Greensand and Chalk of the Wincanton Sheet (297): 1:10 000 quarter sheets ST73SE, ST83NW, SW, SE, ST84SW, ST93NW & SW. *British Geological Survey Technical Report*, WH/94/279R.

WOODS, M A. 1995a. Macrofossils from the Lower Chalk of the Wincanton Sheet (297). *British Geological Survey Technical Report*, WH/95/85R.

WOODS, M A. 1995b. Redetermination of some *Micraster* species from the Upper Chalk of the Wincanton (297) and Salisbury (298) sheets. *British Geological Survey Technical Report*, WH/95/228R.

GEOPHYSICAL DATA

Much of the data held in the archives of the Regional Geophysics Unit, Keyworth, has been synthesised and incorporated in Chapters 3 and 8 and in the various maps listed above.

REMOTE SENSING DATA

Aerial photographs, together with some Landsat imagery, for the district are held at the Exeter Office.

BIBLIOGRAPHY

Some 746 geological references for the area are stored in an ENDNOTE bibliographical database, Exeter.

HYDROGEOLOGY

The older source data is listed above. Additional data is held in the archives of the Hydrogeology Group, Wallingford. Much of these data have been synthesised and incorporated in Chapter 2.

QUARRIES AND MINES

Most quarries are recorded in Chapter 2; additional detail occurs in the Open-file reports listed above.

SSSIs

Geological Sites of Special Scientific Interest (SSSIs) are briefly described in Chapter 2. Further details of all types of SSSIs in the area can be obtained from English Nature through their Taunton (for Somerset) or Devizes (for Wiltshire) offices. There is no SSSI in the Dorset part of the Wincanton district.

MINERAL EXPLORATION

Enquiries relating to bulk-mineral exploration should be sent to the Minerals Group, Keyworth. The Basin Analysis & Stratigraphy Group, Keyworth, hold data on petroleum exploration in the district.

Addresses

British Geological Survey, Keyworth, Nottingham, NG12 5GG
Tel: 0115 936 3100   Fax: 0115 936 3200

British Geological Survey, London Information Office, Natural History Museum, Earth Galleries, Exhibition Road, South Kensington, London, SW7 2DE.
Tel: 0207 589 4090   Fax: 0207 584 8270

British Geological Survey, 30 Pennsylvania Road, Exeter, Devon, EX4 6BX
Tel: 01392 278312   Fax: 01392 437505

Web site   http://www.bgs.ac.uk

REFERENCES

ALLEN, D J, and HOLLOWAY, S. 1984. *The Wessex Basin.* Investigation of the geothermal potential of the UK. (Keyworth, Nottingham: Institute of Geological Sciences.)

ANDREWS, W R, and JUKES-BROWNE, A J. 1894. The Purbeck Beds of the Vale of Wardour. *Quarterly Journal of the Geological Society of London*, Vol. 50, 44–71.

APPLETON, J D, and BALL, T K. 1995. Radon and background radioactivity from natural sources: characteristics, extent and relevance to planning and development in Great Britain. *British Geological Survey Technical Report*, WP/95/2.

ARKELL, W J. 1927. The Corallian rocks of Oxford, Berkshire and north Wiltshire. *Philosophical Transactions of the Royal Society*, Vol. 216 (1928), 67–181.

ARKELL, W J. 1929. A monograph of British Corallian Lamellibranchia. *Monograph of the Palaeontographical Society, London.*

ARKELL, W J. 1933. *The Jurassic System in Great Britain.* (Oxford: Clarendon Press.)

ARKELL, W J. 1947. The geology of the country around Weymouth, Swanage, Corfe and Lulworth. *Memoir of the Geological Survey of Great Britain.* Sheets 341, 342, 343 and parts of 327, 328 and 329 (England and Wales).

ARKELL, W J. 1951–1958. Monograph of the English Bathonian ammonites. *Monograph of the Palaeontographical Society London.*

ATES, A, and KEARY, P. 1993a. Structure of Blackdown Pericline, Mendip Hills from gravity and seismic data. *Journal of the Geological Society of London,* Vol. 150, 729–736.

ATES, A, and KEARY, P. 1993b. Deep structure of the East Mendip Hills from gravity, aeromagnetic and seismic reflection data. *Journal of the Geological Society of London,* Vol. 150, 1055–1063.

AVON and DORSET RIVER AUTHORITY. 1970. *Periodic survey of the water resources and future demand with the Avon and Dorset River Authority.* (Bournemouth: Avon and Dorset River Authority.)

AVON and DORSET RIVER AUTHORITY. 1973. *Upper Wylye investigation.* (Poole: Avon and Dorset River Authority.)

BARKER, D. 1966. Ostracoda from the Portland Beds of Dorset. *Bulletin of the British Museum (Natural History), Geology,* Vol. 11, 447–457.

BARROW, G. 1919. Some future work for the Geologists' Association. *Proceedings of the Geologists' Association,* Vol. 30, 2–48.

BARTLETT, B P, and SCANES, J. 1917. Excursion to Mere and Maiden Bradley in Wiltshire. *Proceedings of the Geologists' Association,* Vol. 27, 117–134.

BARTON, C M. 1990. Geology of the Stalbridge district (Dorset and Somerset). 1:10 000 Sheet ST71NW. *British Geological Survey Technical Report,* WA/90/21.

BARTON, C M. 1994. Geology of the Mere–Maiden Bradley district (Wiltshire and Dorset). 1:10 000 sheets ST83SW and ST83NW. *British Geological Survey Technical Report,* WA/94/87.

BARTON, C M, EVANS, D J, BRISTOW, C R, FRESHNEY, E C, and KIRBY, G A. 1998. Reactivation of relay ramps and structural evolution of the Mere Fault and Wardour Monocline, northern Wessex Basin. *Geological Magazine,* Vol. 135, 383–395.

BARTON, C M, IVIMEY–COOK, H C, LOTT, G K, and TAYLOR, R T. 1993. The Purse Caundle Borehole, Dorset: Stratigraphy and sedimentology of the Inferior Oolite and Fuller's Earth in the Sherborne area of the Wessex Basin. *Research Report of the British Geological Survey, Onshore Geology Series,* SA/93/01.

BATES, S E H. 1905. Penselwood. *Proceedings of the Somerset Archaeological & Natural History Society,* Vol. 50, 50–67.

BLAKE, J F. 1880. On the Portland rocks of England. *Quarterly Journal of the Geological Society of London,* Vol. 36, 189–236.

BLAKE, J F, and HUDLESTON, W H. 1877. On the Corallian rocks of England. *Quarterly Journal of the Geological Society of London,* Vol. 33, 260–405.

BOSWELL, P G H. 1924. The petrography of the Cretaceous and Tertiary outliers of the west of England. *Quarterly Journal of the Geological Society of London,* Vol. 79, 205–230.

BRADSHAW, M J, and CRIPPS, D W. 1992. Mid Callovian. In *Atlas of palaeogeography and lithofacies* Vol. 13, COPE, C J W, INGHAM, J K, and RAWSON, P F. (editor). (London: Geological Society.)

BRISTOW, C R. 1989a. Geology of the East Stour–Shaftesbury district (Dorset). 1:10 000 sheets ST82SW and 82SE. *British Geological Survey Technical Report,* WA/89/58.

BRISTOW, C R. 1989b. Geology of Sheets ST71NE and SE (Marnhull–Sturminster Newton, Dorset). *British Geological Survey Technical Report,* WA/89/59.

BRISTOW, C R. 1990a. Geology of the West Stour–Cucklington district. 1:10 000 sheets ST72NE and 72NE. *British Geological Survey Technical Report,* WA/90/48.

BRISTOW, C R. 1990b. Geology of the Berwick St John district (Wiltshire). 1:10 000 Sheet ST92SW. *British Geological Survey Technical Report,* WA/90/49.

BRISTOW, C R. 1993. Geology of the Gillingham–Semley district, Dorset. 1:10 000 sheets ST82NW (Gillingham) and ST82NE (Semley). *British Geological Survey Technical Report,* WA/93/87.

BRISTOW, C R. 1994. Geology of the Horningsham–Warminster south district (Wiltshire). 1:10 000 sheets ST84SW (Horningsham) and ST84SE (Warminster south). *British Geological Survey Technical Report,* WA/94/88.

BRISTOW, C R. 1995a. Geology of the Brixton Deverill–East Knoyle district (Wiltshire). 1:10 000 sheets ST83NE (Brixton Deverill) and ST83SE (East Knoyle). *British Geological Survey Technical Report,* WA/95/13.

BRISTOW, C R. 1995b. Geology of the Tisbury district (Wiltshire). 1:10 000 Sheet ST92NW (Tisbury) and the Jurassic of Sheet ST93SW. *British Geological Survey Technical Report,* WA/95/82.

BRISTOW, C R. 1995c. Geology of the Great Ridge–Hindon district (Wiltshire). 1:10 000 sheets ST93NW (Great Ridge) and ST93SW (Hindon). *British Geological Survey Technical Report,* WA/95/85.

BRISTOW, C R. 1996. Geology of the Heytesbury district, Wiltshire. 1:10 000 Sheet ST94SW (Heytesbury). *British Geological Survey Technical Report,* WH/95/86.

BRISTOW, C R, and LOTT, G K. 1994. The stratigraphy and building stone potential of the Portland Beds in the western part of the Vale of Wardour. *British Geological Survey Technical Report*, WA/94/97C.

BRISTOW, C R, and OWEN, H G. 1991. A temporary section in the Gault at Fontmell Magna in north Dorset. *Proceedings of the Dorset Natural History and Archaeological Society*, Vol. 112, 95–97.

BRISTOW, C R, and WESTHEAD, R K. 1993. Geology of the Evercreech-Batcombe district (Somerset). 1:10 000 sheets ST63NW (Evercreech) and ST63NE (Batcombe). *British Geological Survey Technical Report*, WA/93/89.

BRISTOW, C R, BARTON, C M, FRESHNEY, E C, WOOD, C J, EVANS, D J, COX, B M, IVIMEY-COOK, H C, and TAYLOR, R T. 1995. Geology of the country around Shaftesbury. *Memoir of the Geological Survey of Great Britain*, Sheet 313 (England and Wales).

BRISTOW, C R, COX, B M, WOODS, M A, PRUDDEN, H C, SOLE, D, EDMUNDS, M, and CALLOMON, J H. 1992. The geology of the A303 trunk road between Wincanton, Somerset and Mere, Wiltshire. *Proceedings of the Dorset Natural History and Archaeological Society*, Vol. 117, 139–143.

BRISTOW, C R, COX, B M, PRUDDEN, H C, CALLOMON, J H, and PAGE, K N. 1993. Temporary sections in the Kellaways and Oxford Clay Formations of the Wincanton area. *Proceedings of the Somerset Archaeological & Natural History Society for 1992*, Vol. 136, 213–220.

BRISTOW, C R, FRESHNEY, E C, and PENN, I E. 1991. The geology of the country around Bournemouth. *Memoir of the Geological Survey of Great Britain*. Sheet 329 (England and Wales).

BRISTOW, C R, MORTIMORE, R N, and WOOD, C J. 1997. Lithostratigraphy for mapping the Chalk of southern England. *Proceedings of the Geologists' Association*, Vol. 108, 293–315.

BROMLEY, R G, and GALE, A S. 1982. The lithostratigraphy of the English Chalk Rock. *Cretaceous Research*, Vol. 3, 273–306.

BROOKS, M, BAYERLY, M, and LLEWELLYN, D J. 1977. A new geological model to explain the gravity gradient across Exmoor, north Devon. *Journal of the Geological Society of London*, Vol. 133, 385–393.

BRYANT, I D, KANTOROWICZ, J D, and LOVE, C F. 1988. The origin and recognition of laterally continuous carbonate-cemented horizons in the Upper Lias Sands of southern England. *Marine and Petroleum Geology*, Vol. 5, 108–133.

BUCKMAN, J. 1879. On the so-called Midford Sands. *Quarterly Journal of the Geological Society of London*, Vol. 35, 736–743.

BUCKMAN, S S. 1889. On the Cotteswold, Midford and Yeovil Sands, and the divisions between the Lias and Oolite. *Quarterly Journal of the Geological Society of London*, Vol. 45, 440–474.

BUCKMAN, S S. 1922. Jurasic chronology: II — Preliminary studies. Certain Jurassic strata near Eypesmouth (Dorset): the Junction-Bed of Watton Cliff and associated rocks. *Quarterly Journal of the Geological Society of London*, Vol. 78, 378–436.

BUCKMAN, S S. 1927. Jurassic chronology: III — Some faunal horizons in Cornbrash. *Quarterly Journal of the Geological Society of London*, Vol. 83, 1–37.

CALLOMON, J H. 1964. Notes on the Callovian and Oxfordian stages. *Comptes rendus et memoires Colloque Jurassique Luxembourg 1962*, 269–291.

CALLOMON, J H, DIETL, G, and PAGE, K N. 1989. On the ammonite faunal horizons and standard zonations of the lower Callovian Stage in Europe. 359–376 in *2nd International Symposium on Jurassic Stratigraphy (Lisboa 1987)*. Vol. 1.

CARTER, D J, and HART, M B. 1977. Aspects of mid-Cretaceous stratigraphical micropalaeontology. *Bulletin of the British Museum (Natural History) (Geology)*, Vol. 29, 1–135.

CHADWICK, R A. 1985. Permian, Mesozoic and Cenozoic structural evolution of England and Wales in relation to the principles of extension and inversion tectonics. 9–25 in *Atlas of onshore sedimentary basins in England and Wales: Post-Carboniferous tectonics and stratigraphy*. WHITTAKER, A (editor). (Glasgow: Blackie.)

CHADWICK, R A. 1986. Extension tectonics in the Wessex Basin, southern England. *Journal of the Geological Society of London*, Vol. 143, 465–488.

CHADWICK, R A. 1993. Aspects of basin inversion in southern Britain. *Journal of the Geological Society of London*, Vol. 150, 311–322.

CHADWICK, R A, KENOLTY, N, and WHITTAKER, A. 1983. Crustal structure beneath southern England from deep seismic reflection profiles. *Journal of the Geological Society of London*, Vol. 140, 893–911.

CHADWICK, R A, and KIRBY, G A. 1982. The geology beneath the Lower Greensand/Gault surface in the Vale of Wardour area. *Report of the Institute of Geological Sciences*, 82/1, 15–18.

CHADWICK, R A, PHAROAH, T C, and SMITH, N J P. 1989. Lower crustal heterogeneity beneath Britain from deep seismic reflection data. *Journal of the Geological Society of London*, Vol. 146, 617–630.

CLARKE, R H, and SOUTHWOOD, T R E. 1989. Risks from ionising radiation. *Nature, London*, Vol. 338, 197–198.

COLTER, V S, and HAVARD, D J. 1981. The Wytch Farm Oil Field, Dorset. 493–503 in *Petroleum geology of the Continental Shelf of North-West Europe*. ILLING, V C, and HOBSON, G D (editors). (London: Institute of Petroleum.)

COPE, J C W, GETTY, T, HOWARTH, M K, MORTON, N, and TORRENS, H S. 1980a. A correlation of Jurassic rocks of the British Isles. Part One: Introduction and Lower Jurassic. *Special Report of the Geological Society of London*, No. 14.

COPE, J C W, DUFF, K L, PARSONS, C F, TORRENS, H S, WIMBLEDON, W A, and WRIGHT, J K. 1980b. A correlation of Jurassic rocks in the British Isles. Part Two: Middle and Upper Jurassic. *Special Report of the Geological Society of London*, No. 15.

COPE, J C W, HALLAM, A, and TORRENS, H S. 1969. *Guide for Dorset and south Somerset: International Field Symposium on the Jurassic, Excursion No. 1*. (Keele: Keele University Geology Department.)

CORNWELL, J D, and CHACKSFIELD, B C. 1997. Geophysical investigations in the Wincanton district. *British Geological Survey Technical Report*, WK/96/10.

COX, B M, and GALLOIS, R W. 1979. Description of the standard stratigraphical sequence of the Upper Kimmeridge Clay, Ampthill Clay and West Walton Beds. *Report of the Institute of Geological Sciences*, No. 78/19, 68–72.

COX, B M, and GALLOIS, R W. 1981. The stratigraphy of the Kimmeridge Clay of the Dorset type area and its correlation with some other Kimmeridgian sequences. *Report of the Institute of Geological Sciences*, No. 80/4.

DAVIES, D K. 1967. Origin of friable sandstone–calcareous sandstone rhythms in the Upper Lias of England. *Journal of Sedimentary Petrology*, Vol. 39, 1344–1370.

DAVIES, D K. 1969. Shelf sedimentation: An example from the Jurassic of Britain. *Journal of Sedimentary Petrology*, Vol. 39, 1344–1370.

DAVIES, D K, ETHERIDGE, F G, and BERG, R R. 1971. Recognition of barrier environments. *Bulletin of the American Association of Petroleum Geologists*, Vol. 55, 550–565.

DE LA BECHE, H T. 1846. On the formation of the rocks of south Wales and south western England. *Memoir of the Geological Survey of Great Britain*, Vol. 1, 1–296.

DELAIR, J B. 1959. The Mesozoic reptiles of Dorset. *Proceedings of the Dorset Natural History and Archaeological Society*, Vol. 80, 52–90.

DELAIR, J B. 1960. The Mesozoic reptiles of Dorset. *Proceedings of the Dorset Natural History and Archaeological Society*, Vol. 81, 59–85.

DIETL, G. 1982. Das wirkliche Fundniveau von *Ammonites aspidoides* Oppel (Ammonoidea, Mittl. Jura) am locus typicus. *Stuttgarter Beiträge zur Naturkunde*, Vol. Series B, No. 87, 13–21.

DONOVAN, D T. 1958. The Lower Lias Section at Cannard's Grave, Shepton Mallet, Somerset. *Proceedings of the Bristol Naturalists Society*, Vol. 29, 393–398.

DONOVAN, D T, BENNETT, R, BRISTOW, C R, CARPENTER, S C, GREEN, G W, HAWKES, C J, PRUDDEN, H C, and STANTON, W I. 1989. Geology of a gas pipeline from Ilchester (Somerset) to Pucklechurch (Avon), 1985. *Proceedings of the Somerset Archaeological & Natural History Society for 1988*, Vol. 132, 297–317.

DONOVAN, D T, and KELLAWAY, G A. 1984. Geology of the Bristol district: the Lower Jurassic rocks. *Memoir of the Geological Survey of Great Britain*.

DOUGLAS, J A, and ARKELL, W J. 1928. The stratigraphical distribution of the Cornbrash. *Quarterly Journal of the Geological Society of London*, Vol. 84, 117–178.

DOUGLAS, J A, and ARKELL, W J. 1932. The stratigraphical distribution of the Cornbrash. II The North-eastern area. *Quarterly Journal of the Geological Society of London*, Vol. 88, 112–170.

DRUMMOND, P V O. 1970. The Mid-Dorset Swell. Evidence of Albian–Cenomanian movements in Wessex. *Proceedings of the Geologists' Association*, Vol. 81, 679–714.

DUFF, K L, MCKIRDY, A P, and HARLEY, M J. 1985. *New sites for old. A student's guide to the geology of the east Mendips.* (Peterborough: Nature Conservancy Council.)

EDMUNDS, F H. 1938. A contribution on the physiography of the Mere district, Wiltshire, with report of field meeting. *Proceedings of the Geologists' Association*, Vol. 49, 174–196.

EDWARDS, W, and PRINGLE, J. 1926. On a borehole in the Lower Oolitic Rocks at Wincanton, Somerset. *Summary of Progress of the Geological Survey for 1925*, 183–188.

EVANS, D J, and CHADWICK, R A. 1994. Basement–cover relationships in the Shaftesbury area of the Wessex Basin, southern England. *Geological Magazine*, Vol. 131, 387–394.

FARMER, D L. 1992. Millstones for medieval manors. *The Agricultural History Review*, Vol. 40, 97–111.

FITTON, W H. 1836. Observations on some of the strata between the Chalk and the Oxford Oolite, in the south-east of England. *Transactions of the Geological Society of London*, Vol. 4, 103–389.

FOSTER, S S D, and MILTON, V A. 1974. The permeability and storage of an unconfined aquifer. *Hydrogeological Sciences Bulletin*, Vol. 19, 485–500.

FRASER, J G. 1863. Descriptions of the Lydgate and of the Buckhorn Weston Railway Tunnels. *Minutes and Proceedings of the Institute of Civil Engineers*, Vol. 22, 371–384.

FRESHNEY, E C. 1994. Geology of the Bruton–Holton district, Somerset. 1:10 000 sheets ST63SE and ST62NE. *British Geological Survey Technical Report*, WA/94/02.

FRESHNEY, E C. 1995a. Geology of the Charlton Musgrove–Wincanton district, Somerset. 1:10 000 sheets ST72NW and ST73SW. *British Geological Survey Technical Report*, WA/95/23.

FRESHNEY, E C. 1995b. Geology of the Kilmington–Zeals district, Wiltshire. 1:10 000 sheets ST73NE and ST73SE. *British Geological Survey Technical Report*, WA/95/24.

FÜRSICH, F T. 1976. The use of macroinvertebrate associations in interpreting Corallian environment. *Palaeogeography, Palaeoclimatology, Palaeoecology*, Vol. 20, 235–256.

GALE, A S. 1989. Field meeting at Folkestone Warren, 29th November, 1987. *Proceedings of the Geologists' Association*, Vol. 100, 73–82.

GALE, A S. 1990. A Milankovitch scale for Cenomanian time. *Terra Nova*, Vol. 1, 420–425.

GALE, A S, WOOD, C J, and BROMLEY, R G. 1987. The lithostratigraphy and marker bed correlation of the White Chalk (late Cenomanian–Campanian) in southern England. *Mesozoic Research*, Vol. 1, 107–118.

GALLOIS, R W. 1979. *Oil shale resources in Great Britain.* (London: Institute of Geological Sciences.)

GALLOIS, R W, and COX, B M. 1976. The stratigraphy of the Lower Kimmeridge Clay of Eastern England. *Proceedings of the Yorkshire Geological Society*, Vol. 41, 13–26.

GALLOIS, R W, and COX, B M. 1977. The stratigraphy of the Middle and Upper Oxfordian sediments of Fenland. *Proceedings of the Geologists' Association*, Vol. 88, 207–228.

GOSTELOW, T P. 1991. Geological processes and their effect on the engineering behaviour of the Gault–UGS: An example from Shaftesbury, Dorset. *British Geological Survey Technical Report*, WN/91/9.

GREEN, G W. 1992. *British regional geology, Bristol and Gloucester region (3rd edition).* (London: HMSO for British Geological Survey.)

GREEN, G W, and WELCH, F B A. 1965. Geology of the country around Wells and Cheddar. *Memoir of the Geological Survey of Great Britain*, Sheet 280 (England and Wales).

GUTMANN, K. 1970. The Corallian Beds at Todber and Whiteway Hill in north Dorset. *Proceedings of the Dorset Natural History and Archaeological Society*, Vol. 91, 123–133.

HALLAM, A. 1970. *Gyrochorte* and other trace fossils in the Forest Marble (Bathonian) of Dorset, England. 189–200 in *Trace fossils*. CRIMES, T P, and HARPER, J C (editors). (Liverpool: Seel House Press.)

HANCOCK, N J. 1982. Stratigraphy, palaeogeography and structure of the east Mendips Silurian inlier. *Proceedings of the Geologists' Association*, Vol. 93, 247–261.

HANTZPERGUE, P. 1989. *Les ammonites kimméridgiennes du haut-found d'Europe occidentale: biochronologie, systématique, évolution, paléobiogéographie.* (Paris.)

HENDERSON, A S, OLIVER, G M, and HART, M B. 1994. The lithostratigraphy of the Mid-Jurassic of north Dorset; preliminary results from three new boreholes. *Proceedings of the Ussher Society*, Vol. 8, 325–327.

HODGSON, J M, CATT, J A, and WEIR, A H. 1967. The origin and development of Clay-with-flints and associated soil horizons on the South Downs. *Journal of Soil Science*, Vol. 18, 85–102.

HOLLINGWORTH, N T J, WARD, D J, SIMMS, M J, and CLOTHIER, P. 1990. A temporary exposure of Lower Lias (Late Sinemurian) at Dimmer Camp, Castle Cary, Somerset, south-west England. *Mesozoic Research*, Vol. 2, 163–180.

HOLLOWAY, S. 1982. Bruton No. 1. Geological Well Completion Report. *Institute of Geological Sciences, Deep Geology Unit*, Report No. 82/8.

HOLLOWAY, S. 1985. Lower Jurassic: the Lias. 37–40 in *Atlas of onshore sedimentary basins in England and Wales: Post-Carboniferous tectonics and stratigraphy*. WHITTAKER, A (editor). (Glasgow & London: Blackie.)

HOLLOWAY, S, and CHADWICK, R A. 1984. The IGS Bruton Borehole (Somerset, England) and its regional structural significance. *Proceedings of the Geologists' Association*, Vol. 95, 165–174.

HOUNSLOW, M W. 1987. Magnetic fabric characteristics of bioturbated wave-produced grain orientation in the Bridport–Yeovil Sands (Lower Jurassic). *Sedimentology*, Vol. 34, 117–128.

HOWE, C. 1983. *Gylla's Hometown*. (Gillingham: Gylla Publishing.)

HUDLESTON, W H. 1881. On the geology of the Vale of Wardour. *Proceedings of the Geologists' Association*, Vol. 7, 161–185.

HUDSON, J D, and MARTILL, D M. 1991. The Lower Oxford Clay: production and preservation of organic matter in the Callovian (Jurassic) of Central England. 363–379 in Modern and ancient continental shelf anoxia. TYSON, R V, and PEARSON, T H. (editors). *Special Publication of the Geological Society of London*, No. 58.

HUDSON, J D, and MARTILL, D M. 1994. The Peterborough Member (Callovian, Middle Jurassic) of the Oxford Clay Formation at Peterborough, UK. *Journal of the Geological Society of London*, Vol. 151, 113–124.

HULL, E, and WHITAKER, W. 1861. The geology of parts of Oxfordshire and Berkshire. *Memoir of the Geological Survey of Great Britain*, Sheet 13 (England and Wales).

HUTCHINSON, J N. 1981. Damage to slopes produced by seepage erosion on sands (abstract). in *Water-related exogenous geological processes and prevention of their negative impact on the environment*. (USSR: Alma Ata.)

INSTITUTE OF GEOLOGICAL SCIENCES. 1979. *Hydrogeological map of the Chalk and associated minor aquifers of Wessex*. (London: Institute of Geological Sciences.)

JEFFERIES, R P S. 1963. The stratigraphy of the *Actinocamax plenus* Subzone (Turonian) in the Anglo-Paris Basin. *Proceedings of the Geologists' Association*, Vol. 74, 1–33.

JENKYNS, H C, and SENIOR, J R. 1991. Geological evidence for intra-Jurassic faulting in the Wessex Basin and its margins. *Journal of the Geological Society of London*, Vol. 148, 245–260.

JUKES-BROWNE, A J, and HILL, W. 1900. The Cretaceous rocks of Britain. 1. The Gault and Upper Greensand. *Memoir of the Geological Survey of Great Britain*.

JUKES-BROWNE, A J, and HILL, W. 1903. The Cretaceous rocks of Britain. 2. The Lower and Middle Chalk of England. *Memoir of the Geological Survey of Great Britain*.

JUKES-BROWNE, A J, and HILL, W. 1904. The Cretaceous rocks of Britain. 4. Upper Chalk of England. *Memoir of the Geological Survey of Great Britain*.

JUKES-BROWNE, A J, and SCANES, J. 1901. The Upper Greensand and Chloritic Marl of Mere and Maiden Bradley. *Quarterly Journal of the Geological Society of London*, Vol. 57, 96–125.

KANTOROWICZ, J D, BRYANT, I D, and DAWANS, J M. 1987. Controls on the geometry and distribution of carbonate cements in Jurassic sandstones: Bridport Sands, southern England and Viking Group, Troll Field, Norway. *In* Diagenesis of sedimentary sequences. MARSHALL, J D (editor). *Geological Society of London Special Publication*, No. 36.

KELLAWAY, G A, and HANCOCK, P L. 1983. Structure of the Bristol district, the Forest of Dean and the Malvern fault zone. 74–87 in *The Variscan Fold Belt in the British Isles*. HANCOCK, P L (editor). (Bristol: Adam Hilger Ltd.)

KELLAWAY, G A, and WELCH, F B A. 1948. *British regional geology: Bristol and Gloucester district (2nd edition)*. (London: HMSO for Institute of Geological Sciences.)

KELLAWAY, G A, and WELCH, F B A. 1993. Geology of the Bristol district. *Memoir of the Geological Survey of Great Britain*. Bristol Special Sheet (England and Wales).

KELLAWAY, G A, and WILSON, V. 1941. An outline of the geology of Yeovil, Sherborne and Sparkford Vale. *Proceedings of the Geologists' Association*, Vol. 52, 131–174.

KENNEDY, W J. 1970. A correlation of the uppermost Albian and the Cenomanian of south-west England. *Proceedings of the Geologists' Association*, Vol. 81, 613–677.

KENOLTY, N, CHADWICK, R A, BLUNDELL, D J, and BACON, M. 1981. Deep seismic reflection survey over the Variscan Front of southern England. *Nature, London*, Vol. 293, 451–453.

KNOX, R W O. 1982. Clay mineral trends in cored Lower and Middle Jurassic sediments in the Winterborne Kingston borehole, Dorset. 91–96 *in* The Winterborne Kingston borehole, Dorset, England. RHYS, G H, LOTT, G K, and CALVER, M A (editors). *Report of the Institute of Geological Sciences*, No. 81/3.

LOTT, G K. 1993. Thin section petrography of Upper Jurassic (Corallian Oolite Fm/Kimmeridge Clay Fm/Portland 'Stone') sediments in the Salisbury Sheet (298) area, Wiltshire. *British Geological Survey Technical Report*, WH/93/81.

LOTT, G K. 1994. The petrology of sandstones and siltstones from the Portland Sand Formation, Tisbury Borehole, Wiltshire. *British Geological Survey Technical Report*, WH/94/178R.

LOTT, G K, and STRONG, G E. 1982. The petrology and petrography of the Sherwood Sandstone (?Middle Triassic) of the Winterborne Kingston borehole, Dorset. *In* The Winterborne Kingston borehole, Dorset, England. RHYS, G H, LOTT, G K, and CALVER, M A (editors). *Report of the Institute of Geological Sciences*, No. 81/3.

LOVEDAY, J. 1962. Plateau deposits on the Southern Chiltern Hills. *Proceedings of the Geologists' Association*, Vol. 73, 83–102.

MACDONALD, A M, and COLEBY, L M. in press. Chalk. In *The physical properties of major aquifers in England and Wales*. ALLEN, D J, and ROBINSON, V K (editors). (National Rivers Authority and British Geological Survey.)

MACQUAKER, J H S. 1994. A lithofacies study of the Peterborough Member, Oxford Clay Formation (Jurassic), UK: an example of sediment bypass in a mudstone succession. *Journal of the Geological Society of London*, Vol. 151, 161–172.

MANNERS, J E. 1971. The stonemason of Tucking Mill. *Country Life*, Vol. 150(3880), 1082–1084.

MANSELL-PLEYDELL, J C. 1890. Memoir upon a new Ichthyoptergian from the Kimmeridge Clay of Gillingham, Dorset, *Opthalmosaurus Pleydelli*. *Proceedings of the Dorset Natural History and Archaeological Society*, Vol. 11, 7–15.

MCKENZIE, D. 1978. Some remarks on the development of sedimentary basins. *Journal of the Geological Society of London*, Vol. 150, 25–32.

MITCHELL, C J, and MURPHY, H A. 1995. Clay mineralogy of Frome Clay from Lodge Farm and Bratton Seymour, Somerset. *British Geological Survey Technical Report* MPSR/95/21.

MORTER, A A, and WOOD, C J. 1983. The biostratigraphy of Upper Albian–Lower Cenomanian *Aucellina* in Europe. *Zitteliana*, Vol. 10, 515–529.

MORTIMORE, R N. 1983. The stratigraphy and sedimentation of the Turonian Campanian in the southern province of England. *Zitteliana*, Vol. 10, 22–41.

MORTIMORE, R N. 1986. Stratigraphy of the Upper Cretaceous White Chalk of Sussex. *Proceedings of the Geologists' Association*, Vol. 98, 97–139.

MORTIMORE, R N. 1987. Upper Cretaceous Chalk in the North and South Downs, England: a correlation. *Proceedings of the Geologists' Association*, Vol. 98, 77-86.

MORTIMORE, R N, and POMEROL, B. 1987. Correlation of the Upper Cretaceous White Chalk (Turonian to Campanian) in the Anglo-Paris Basin. *Proceedings of the Geologists' Association*, Vol. 98, 97–143.

MORTIMORE, R N, and WOOD, C J. 1986. The distribution of flint in the English Chalk, with particular reference to the 'Brandon Flint Series' and the high Turonian flint maximum. in *The scientific study of flint and chert; papers from the Fourth International Flint Symposium*, Vol. 1. SIEVEKING, G D G, and HART, M B (editors). (Cambridge: Cambridge University Press.)

MORTON, A C. 1982. Heavy minerals from the sandstones of the Winterborne Kingston borehole, Dorset. in *The Winterborne Kingston borehole, Dorset, England*. RHYS, G H, LOTT, G K, and CALVER, M A (editors). *Report of the Institute of Geological Sciences*, No. 81/3.

MOTTRAM, B H. 1957. Whitsun field meeting at Shaftesbury. *Proceedings of the Geologists' Association*, Vol. 67 (for 1956), 160–167.

MOTTRAM, B H. 1961. Contributions to the geology of the Mere Fault and the Vale of Wardour Anticline. *Proceedings of the Geologist' Association*, Vol. 72, 187–203.

OWEN, H G. 1971. Middle Albian stratigraphy in the Anglo-Paris Basin. *Bulletin of the British Museum (Natural History) Geology*, Vol. Supplement 8.

OWEN, H G. 1976. The stratigraphy of the Gault and Upper Greensand of the Weald. *Proceedings of the Geologists' Association*, Vol. 86 (for 1975), 475–498.

PAGE, K N. 1988. The stratigraphy and ammonites of the British Lower Callovian. PhD thesis, University College, London.

PAGE, K N. 1989. A stratigraphical revision for the English Lower Callovian. *Proceedings of the Geologists' Association*, Vol. 100, 363–382.

PAGE, K N. 1995. Biohorizons and zonules: intra-subzonal units in Jurassic ammonite stratigraphy. *Palaeontology*, Vol. 38, 801–814.

PAGE, K N. 1996. Observations on the succession of ammonite faunas in the Bathonian (Middle Jurassic) of south-west England, and their correlation with a Sub-Mediterranean 'Standard Zonation'. *Proceedings of the Ussher Society*, Vol. 9, 045–053.

PARSONS, C F. 1975. Ammonites from the Doulting Conglomerate Bed (Upper Bajocian, Jurassic) of Somerset. *Palaeontology*, Vol. 18, 191–205.

PARSONS, C F. 1979. A stratigraphic revision of the Inferior Oolite of Dundry Hill, Bristol. *Proceedings of the Geologists' Association*, Vol. 90, 133–151.

PENN, I E. 1982. Middle Jurassic stratigraphy and correlation of the Winterborne Kingston Borehole, Dorset. *Report of the Institute of Geological Sciences*, No. 81/3.

PENN, I E, and WYATT, R J. 1979. The Bathonian strata of the Bath–Frome area 2. The stratigraphy and correlation of the Bathonian strata in the Bath–Frome area. *Report of the Institute of Geological Sciences*, No. 78/22.

PRICE, M, MORRIS, B L, and ROBERTSON, A S. 1982. Recharge mechanisms and groundwater flow in the Chalk and Permian aquifers using double packer injection testing. *Journal of Hydrology*, Vol. 54, 401–423.

PRINGLE, J. 1910. On a boring at Stowell, Somerset. *Geological Survey of Great Britain, Summary of Progress for 1909*, 68–70.

REID, C. 1899. The geology of the country around Dorchester. *Memoir of the Geological Survey of Great Britain*. Sheet 328 (England and Wales).

REID, C. 1903. Geology of the country around Salisbury. *Memoir of the Geological Survey of Great Britain*. Sheet 298 (England and Wales).

RICHARDSON, L. 1906. On a section of Middle and Upper Lias rocks near Evercreech, Somerset. *Geological Magazine*, Vol. 3, 368–369.

RICHARDSON, L. 1907. The Inferior Oolite and contiguous deposits of the Bath–Doulting district. *Quarterly Journal of the Geological Society of London*, Vol. 63, 383–436.

RICHARDSON, L. 1909a. Excursion to the Frome district, Somerset. *Proceedings of the Geologists' Association*, Vol. 21, 209–228.

RICHARDSON, L. 1909b. On some Middle and Upper-Lias sections near Batcombe, Somerset. *Geological Magazine*, Vol. 6, 540–542.

RICHARDSON, L. 1911. The Rhaetic and contiguous deposits of west, mid and part of east Somerset. *Quarterly Journal of the Geological Society of London*, Vol. 67, 1–74.

RICHARDSON, L. 1916. The Inferior Oolite and contiguous deposits of the Doulting–Milborne Port district (Somerset). *Quarterly Journal of the Geological Society of London*, Vol. 71, 473–520.

ROBINSON, N D. 1986. Lithostratigraphy of the Chalk Group of the North Downs, southeast England. *Proceedings of the Geologists' Association*, Vol. 97, 141–170.

ROSS, M S. 1985. Kington Magna: a parish survey. *Proceedings of the Dorset Natural History and Archaeological Society*, Vol. 107, 89–102.

ROSS, M S. 1992. Brickmaking at Gillingham and Motcombe, Dorset. *Proceedings of the Dorset Natural History and Archaeological Society*, Vol. 113 (for 1991), 17–22.

SAVAGE, R J G. 1977. 9. The Mesozoic strata of the Mendip Hills. 85–100 in *Geological excursions in the Bristol district*. SAVAGE, R J G (editor). (Bristol: University of Bristol.)

SEELEY, H G. 1869. *Index to the fossil remains of Aves, Ornithosauria and Reptilea from the Secondary System of strata, arranged in the Woodwardian Museum of The University of Cambridge*. (Cambridge: Cambridge University Press.)

SHEPPARD, L M. 1981. Bathonian ostracod correlation north and south of the English Channel with description of two new Bathonian ostracods. 73–89 in *Microfossils from recent and fossil shelf seas*. NEALE, J W, and BRASIER, M D (editors). (Chichester: Ellis Horwood.)

SMART, J G O. 1955. Notes on the geology of the Alton Pancras district, Dorset. *Bulletin of the Geological Survey of Great Britain*, No. 9, 42–49.

SMART, J G O, BISSON, G, and WORSSAM, B C. 1966. Geology of the country around Canterbury and Folkestone. *Memoir of the*

Geological Survey of Great Britain, Sheets 289, 305 and 306 (England and Wales).

SPATH, L F. 1923–1943. A monograph of the Ammonoidea of the Gault. *Monograph of the Palaeontological Society*, Vol. 1, pts 1–7 (1923–1930); Vol. 2, pts 8–16 (1931–1943).

SUN, S Q. 1989. A new interpretation of the Corallian (Upper Jurassic) cycles of the Dorset coast, southern England. *Geological Journal*, Vol. 24, 139–158.

SYLVESTER–BRADLEY, P C, and HODSON, F. 1957. The Fuller's Earth of Whatley, Somerset. *Geological Magazine*, Vol. 94, 312–325.

TALBOT, M R. 1973. Major sedimentary cycles in the Corallian beds (Oxfordian) of southern England. *Palaeogeography, Palaeoclimatology, Palaeoecology*, Vol. 14, 293–317.

TAYLOR, R T. 1990. Geology of the Charlton Horethorne district (Somerset and Dorset). 1:10 000 Sheet ST62SE. *British Geological Survey Technical Report*, WA/89/69.

TAYLOR, R T. 1991. Geology of the Templecombe district (Somerset). 1:10 000 sheets ST72SW. 1:10 000 Sheet ST72SW. *British Geological Survey Technical Report*, WA/91/31.

TORRENS, H S. 1969a. International Field Symposium on the British Jurassic. A1–A71 in *Excursion No. 1. Guide for Dorset and south Somerset*.

TORRENS, H S. 1969b. Field meeting in the Sherborne–Yeovil district. *Proceedings of the Geologists' Association*, Vol. 80, 301–324.

TORRENS, H S. 1974. Standard zones of the Bathonian. *Memoires Bureau de Recherches Géologiques et Minières*, Vol. 75, 581–604.

TORRENS, H S. 1980. Bathonian correlation chart. *Geological Society of London Special Report*, No.15, 21–44.

TRESISE, G R. 1960. Aspects of the Lithology of the Wessex Upper Greensand. *Proceedings of the Geologists' Association*, Vol. 71, 316–339.

TRESISE, G R. 1961. The nature and origin of chert in the Upper Greensand of Wessex. *Proceedings of the Geologists' Association*, Vol. 72, 333–356.

TUTCHER, J W, and TRUEMAN, A E. 1925. The Liassic rocks of the Radstock District (Somerset). *Quarterly Journal of the Geological Society of London*, Vol. 81, 595–666.

WESTHEAD, R K. 1994. Geology of the Chesterblade–Wanstrow district (Somerset). 1:10 000 sheets ST64SE (Chesterblade) and ST74SW (Wanstrow). *British Geological Survey Technical Report*, WA/93/90.

WESTHEAD, R K. 1995. Geology of the Brewham–Trudoxhill district (Somerset and Wiltshire). 1:10 000 sheets ST73NW and 74SE. *British Geological Survey Technical Report*, WA/95/10.

WHITAKER, W. 1864. Geology of parts of Middlesex, Hertfordshire, etc. *Memoir of the Geological Survey of Great Britain*, Sheet 7.

WHITAKER, W, and EDWARDS, W. 1926. Wells and springs of Dorset. *Memoir of the Geological Survey of Great Britain.*

WHITE, H G O. 1923. Geology of the country south and west of Shaftesbury. *Memoir of the Geological Survey of Great Britain*, Sheet 313 (England and Wales).

WILLIAMS, G D, and CHAPMAN, T J. 1986. The Bristol–Mendip foreland thrust belt. *Quarterly Journal of the Geological Society of London*, Vol. 143, 63–73.

WILSON, R C L. 1968a. Upper Oxfordian palaeogeography of southern England. *Palaeogeography, Palaeoclimatology, Palaeoecology*, Vol. 4, 5–28.

WILSON, R C L. 1968b. Carbonate facies variation within the Osmington Oolite Series in southern England. *Palaeogeography, Palaeoclimatology, Palaeoecology*, Vol. 4, 89–123.

WILSON, V, WELCH, F B A, ROBBIE, J A, and GREEN, G W. 1958. Geology of the country around Bridport and Yeovil. *Memoir of the Geological Survey of Great Britain*, Sheet 312 and 327 (England and Wales).

WIMBLEDON, W A. 1976. The Portland Beds (Upper Jurassic) of Wiltshire. *The Wiltshire Archaeological and Natural History Journal*, Vol. 71, 3–11.

WIMBLEDON, W A. 1980. Portlandian correlation chart. *Special Report of the Geological Society of London*, No. 15, 85–93.

WINWOOD, H H. 1885. The results of further excavations at Pen Pits. *Proceedings of the Somerset Archaeological & Natural History Society*, Vol. 30, 149–152.

WOOD, C J. 1990. The stratigraphy of the Lower Chalk of the Chilterns: observations on the Barton, Sundon, Houghton Regis and Totternhoe quarries. *British Geological Survey Technical Report*, WH/90/397R.

WOODS, M A, and BRISTOW, C R. 1995. A biostratigraphical review of the Gault, Upper Greensand and Chalk of the Wincanton (297) district, Wiltshire. *British Geological Survey Technical Report*, WA/95/60.

WOODWARD, H B. 1888. Further note on the Midford Sands. *Geological Magazine*, Vol. 25, 470.

WOODWARD, H B. 1893. Excursion to Bath, Midford, and Dundry Hill in Somerset, and to Bradford-on-Avon and Westbury, in Wiltshire. *Proceedings of the Geologists' Association*, Vol. 13, 137–139.

WOODWARD, H B. 1895. The Jurassic rocks of Britain. Vol. 5. The Middle and Upper Oolitic rocks of England (Yorkshire excepted). *Memoir of the Geological Survey of Great Britain*.

WRIGHT, C W, and KENNEDY, W J. 1984. The Ammonoidea of the Lower Chalk: Part 1. *Monograph of the Palaeontographical Society*.

WRIGHT, J K. 1980. Oxfordian correlation chart. 61–76 in A correlation of Jurassic rocks in the British Isles. Part two: Middle and Upper Jurassic. COPE, C J W (editor). *Geological Society of London Special Report*, No. 15.

WRIGHT, J K. 1981. The Corallian rocks of north Dorset. *Proceedings of the Geologists' Association*, Vol. 92, 17–32.

WRIGHT, J K. 1982. Reply by the author [to Talbot, G R. 1982. The Corallian rocks of north Dorset]. *Proceedings of the Geologists' Association*, Vol. 93, 312–313.

WRIGHT, J K. 1985. A new exposure of the Corallian Beds in north Dorset. *Proceedings of the Dorset Natural History and Archaeological Society*, Vol. 106 (for 1984), 168.

WRIGHT, J K. 1986. A new look at the stratigraphy, sedimentology and ammonite fauna of the Corallian Group (Oxfordian) of south Dorset. *Proceedings of the Geologists' Association*, Vol. 97, 1–21.

WRIGHT, T. 1856. On the palaeontological and stratigraphical relations of the so-called 'Sands of the Inferior Oolite'. *Quarterly Journal of the Geological Society of London*, Vol. 12, 292–325.

WRIGHT, T. 1860. On the subdivisions of the Inferior Oolite in the south of England, compared with the equivalent beds of that formation in Yorkshire. *Quarterly Journal of the Geological Society of London*, Vol. 16, 1–48.

YOUNG, D. 1972. Brickmaking in Dorset. *Proceedings of the Dorset Natural History and Archaeological Society*, Vol. 93, 213–242.

FOSSIL INDEX

Latinised names only are listed. Species are listed alphabetically regardless of any qualification (aff. cf. etc.).

GENERAL INDEX